the
UNIVERSITY
of
GREENWICH

Food and Agriculture Organization of the United Nations

Preface

Can there be anything left to say on the subject of sustainable forestry? So much has been published on the topic in recent years. Much of the body of literature on the subject has been written in technical terms for an audience of specialists who are not only foresters but also ecologists, sociologists, economists and members of that all-embracing category, the environmentalists. Instead, this book is aimed primarily at a non-technical audience, including decision-makers and the concerned general public. It will, however, be of value to those foresters whose professional education was completed before the concept of sustainability was expanded beyond the sustained supply of timber to include all of the goods and services provided by the forest.

The idea for this book arose from the Paris Declaration, which was issued at the end of the Tenth World Forestry Congress (1991). The declaration called on the world's decision-makers to raise awareness and inform the public so that forest issues could be better understood and appreciated. One of the issues identified by the Congress as being of key importance for forestry as a whole was the concept of sustainable management.

Greater public awareness by itself will not result in the management of forests on a sustainable basis. The public has to be involved in the debate and decisions on the development of systems of management for all types of land use. Pressure from an informed public can make a major contribution to policy formulation and political commitment. The development of national forest policies and the involvement of people was discussed in *Foresters and politics* at the Ninth World Forestry Congress (Mexico City) in 1985.[1] The Stockholm Conference on the Human Environment of 1972, the Seventh World Forestry Congress of 1972 (Buenos Aires) and the Eighth World Forestry Congress of 1978 (Jakarta) have all implicitly or explicitly called upon governments to encourage the sustainable management of forests.

All congresses since the eighth have emphasized the need to involve people in forest policy development.

The requirements for sustainable forest management include not only the involvement of people but also the availability of appropriate techniques and adequate finance. In addition, ways must be found to solve or alleviate the many economic and social problems which, although arising outside the forests, have major impacts on the forest resource. Far-reaching changes in the political environment mean that now there is a chance that people in many more countries may be given the opportunity to participate in forest ownership and management. We believe that sustainable forest management is technically possible, although some of the techniques still require improvement and refining.

Forestry is still a low priority in most countries, and when budgets are drawn up, funds are generally still in very short supply. There are now, however, new and improved methods of assessing the values of services provided by forests, which coupled with an increasing awareness of the indirect benefits they bestow should greatly improve the prospects of meeting the funding needs for sustainable forest management.

We all recognize that the loss and degradation of the world's forests could have far-reaching consequences for humanity. This book is a contribution not only to increasing public awareness of the issues involved but also eventually to the implementation of sustainable forest management and of sustainable land use.

C. H. Murray
Assistant Director-General
Forestry Department
FAO

[1]Paper by J. Westoby to the Ninth World Forestry Congress, Mexico City, July 1985; first published in *Commonwealth Forestry Review*, 64 (2): 105–116.

Contents

Acknowledgements

The work of Gerald Foley, who prepared successive drafts of the text of this book, is gratefully acknowledged.

Thanks are due to Miriam Ostria of The Nature Conservancy (Latin America Program), who provided figures on debt-for-nature-swaps, and to Dr David Wood, who provided information on the valuing of biodiversity.

Inputs to this document and reviews of the text were made by the following FAO staff members: Philippe Alirol, Jim Ball, Jim Bourke, Susan Braatz, C. Chandrasekharan, Gil Child, Mafa Chipeta, Bill Ciesla, Louis Deherve, Dennis Dykstra, Stephen Dembner, Marilyn Hoskins, Klaus Janz, Michel Malagnoux, Tage Michaelsen, Marc de Montalembert, Rudi Heinrich, Jean Paul Lanly, Claude Léger, Soren Lund, Tage Michaelsen, C. H. Murray, S. Muttiah, Francis Ng, Christel Palmberg-Lerche, Christophe Racaut, El Hadji Sène, Wim Sombroek, Paul Vantomme, Philip Wardle and others. Editing was by Jim Ball.

Grateful acknowledgement is made to the authors of the technical papers specially commissioned by FAO that form the foundation of this book. These papers, which will be published shortly by the FAO Forestry Department, are as follows.

■ *On sustainable forest management (with wood production as a main focus and including wood for energy)*
- *Sustainable management of tropical moist forest for wood*, by A. J. Leslie.
- *Aménagement des forêts claires et des savanes en zone soudano-sahélienne (Management of woodlands and savannas in the Sudanian-Sahelian zone)*, by M. Soto Flandez and K. Ouedraogo.
- *Sustainable management of plantation forest in the tropics and subtropics*, by E. Campinhos Jr.

■ *On managing forests for non-wood forest products*
- *Sustainable management of forests in the tropics and subtropics for non-wood forest products*, by G. Wickens.

■ *On management for soil and water conservation*

- *Management for soil and water conservation*, by T. Michaelsen.

■ **On wildlife management**

- *Ensuring sustainable management of wildlife resources: the case of Africa*, by S. S. Ajayi.

■ **On conserving forest biodiversity**

- *Conserving genetic resources in forest ecosystems*, by R. Kemp and C. Palmberg-Lerche.

■ **On sustainable forest management, protection and climate change**

- *Climate change and sustainable forest management*, by D. C. Maciver.
- *Ensuring sustainability of forests through protection from fire, insects and disease*, by W. M. Ciesla.

■ **On policy, institutional and legal aspects of sustainable forest management**

- *Policy, legal and institutional aspects of sustainable forest management*, by M. R. de Montalembert and F. Schmithüsen.

■ **On socio-economic aspects**

- *People's participation in forest and tree management*, by M. Hoskins.

■ **On research for sustainable forest management**

- *Research for sustainable forest management*, by Salleh Mohamed Nor and F. Ng.

■ **On national policies, programmes and experiences in sustainable forest management – case profiles**

- *Bases para el desarrollo forestal sustentable en Chile y tareas pendientes (The basis and action required for sustainable forestry development in Chile)*, by Juan Franco de la Jara.
- *National policies, programmes and experience in sustainable forest management: the case of Indonesia*, by Daryadi Lukito.
- *Sweden: using the forest as a renewable resource*, by R. Hagglund.
- *La gestion forestière durable: la conservation et le développement durable des forêts en Europe (Sustainable forest conservation and development in Europe: the example of France)*, by J. Gadant.
- *New perspectives for managing the United States national forest system*, by H. Salwasser, D. W. MacCleary and T. A. Snellgrove.

CHAPTER 1

Sustainability in a changing world

A world without forests is unthinkable. Yet the world's forests are disappearing at an increasing pace. More forest was lost between 1981 and 1990 than has been recorded in any other decade in human history.

This is not happening for trivial reasons. Forests are being cleared to provide land for food and cash crops. Fuelwood is the main cooking fuel of nearly half the world's people. Wood is essential for building construction and a host of other uses. Timber exports are a source of foreign exchange for many countries.

Cutting trees and clearing forests make perfect sense to those who are doing it. But as the trees disappear there are also losers. Forest dwellers, often the poorest and most vulnerable members of society, are deprived of their homes and livelihoods. Fuelwood and other forest products become harder to obtain. Land is eroded and lakes and dams are filled with silt. With fewer trees to soak up carbon dioxide from the atmosphere, the risk of global warming is increased. As plant and wildlife species become extinct, biological diversity is reduced.

The evolving concept of sustainability

If sustainable forest management means an attempt to freeze the world as it is, it is clearly impossible. There is no prospect of an early end to the pressures causing the clearing of forests. Large areas of forest land, especially in the tropics, will inevitably be converted to agricultural use in the coming decades. Logging and cutting for fuelwood will continue.

The challenge is not to prevent these activities but to manage them. The aim must be to ensure that wood and other forest products are harvested sustainably, that forests are cleared only in a planned and controlled way and that the subsequent land uses are productive and sustainable.

The World Commission on Environment and Development (1987), commonly called the Brundtland Commission, clearly recognized the necessity for a broad approach to sustainability. It said "sustainable development is a process of change in which the exploitation of resources, the direction of investments, the orientation of technological and institutional change are all in harmony and enhance both current and future potential to meet human needs and aspirations".

The concept of sustainability has deep historical roots in forestry. Much early forestry in Europe was concerned with the preservation of forests as wildlife reserves for hunting by kings and nobles. Later came the concept of managing forests for a sustainable yield of timber. This was achieved by balancing the volume to be harvested against the growth predicted from regeneration and planting. "The sustainable management of forests for the production of wood is based on a deceptively simple principle. All that needs to be done is to harvest the wood at an average annual rate no greater than the forest in question can grow it..." (FAO, 1993d).

Management of the forest to provide a sustained yield of timber is still what many foresters have in mind when they talk of sustainable forest management. This definition focuses on the production of wood and does not address the wider issues of the ecological and social functions of forests, with which timber production may only incidentally be compatible or may even conflict. Over the past two decades management solely for wood production has been a cause of steadily growing concern to those affected by the loss of other benefits. It has led, in an increasing number of areas, to confrontation and even physical conflict between loggers and people living in and around the forest areas being harvested.

The concept of sustainable forest management has therefore evolved to encompass these wider issues and values. It is now seen as the multipurpose management of the forest so that its overall capacity to provide goods and services is not diminished. A forest managed in this way will provide timber on a sustainable basis and will continue to provide fuelwood, food and other goods and services for those living in and around it. Its role in the preservation of genetic resources and biological diversity as well as in the protection of the environment will also be maintained.[1]

The application of forest management in this wider sense would, today, represent enormous progress in most of the forest areas of the world. Yet, arguably, even such a broad definition of sustainable forest management is still too restrictive. It could be interpreted as confining forest management to those areas where a sustained yield of forest products and services is, at least in principle, achievable. There is a need for explicit attention to those huge areas where the forests are disappearing as a result of encroachment and clearing for agriculture, where excessive grazing is preventing natural regeneration of trees or where cutting for charcoal-making and fuelwood is leading to the degradation or disappearance of woodlands.

In many countries, it is precisely in such areas that the need for management of the remaining natural forest is greatest. Because of demographic and economic pressures born of poverty, much of this forest clearing and damage is, in practical terms, unavoidable, but a great deal of it is carried out in ways that are needlessly destructive and self-defeating. Legal and social systems often encourage excessive land clearing; trees are cleared from lands that cannot sustain agriculture; the cleared timber is burnt rather than used productively; and long-term social and environmental impacts are ignored.

If sustainable forest management is to be achieved such issues must be addressed, and the concept needs to be widened further – or, perhaps better, viewed as management for sustainability. This approach, of necessity, must be a holistic one, encompassing land-use planning and the wider questions of rural development. Within this framework, it becomes a broad-based and multifaceted activity. Management for sustainability will therefore first be concerned with securing an improved livelihood for the present generation, while maintaining the potential of the forest heritage for future generations. Second, the forest potential must be seen within the broader context of rural development, in which the allocation of land to different uses is part of a dynamic process but where a balance is maintained between forests and other forms of land use in which trees have a role. Third, responsibilities for forest management must be clearly identified and competing interests must be reconciled through dialogue and partnership. Finally, forestry activities will have to compete for scarce financial resources and both the production and the environmental functions must be shown to be worthwhile to both users and financers (FAO, 1993e).

Sustainable forest management therefore involves planning the production of wood for commercial purposes as well as meeting local needs for fuelwood, poles, food, fodder and other purposes. It includes the protection or setting aside of areas to be managed as plant or wildlife reserves or for recreational or environmental purposes. It is concerned with ensuring that conversion of forest lands to agriculture and other uses is done in a properly planned and controlled way. It also covers the regeneration of wastelands and degraded forests, the integration of trees in the farming landscape and the promotion of agroforestry. It is a multidisciplinary task, requiring collaboration among government agencies, non-governmental organizations (NGOs) and, above all, people, especially rural people. It applies at local, national, regional and global levels.

There is a long way to go before such comprehensive forest management is likely to be practised in most parts of the world, particularly where the forests are now under greatest threat. But the concept nevertheless serves to provide an overall policy direction and long-term goal. In the meantime, there are opportunities for intervention towards the object of sustainable management virtually everywhere. What is required is an open-minded, objective assessment of what should and, equally importantly, can be done in any given location followed by mobilization of the will and the resources to carry it out.

An historical perspective

The present threat to forests is not unique in human history, except in its scale. Massive forest clearances were a feature of the early civilizations of Assyria, Babylon, China, Egypt, Greece and Rome. Much of this forest clearing was to provide land for agriculture, but these societies were also voracious consumers of wood for cooking, heating, copper smelting, pottery making, brick-firing, house construction and shipbuilding.

For more than 10 000 years the forests of the Mediterranean region have been cleared; those that remain cover about one-sixth of the region, on land that cannot readily be cultivated, and they have been degraded by humans and their animals. The export of high-quality cedar wood from Lebanon to Egypt began nearly 5 000 years ago and led, eventually, to the virtually complete destruction of the cedar forests. Early commentators were aware of the environmental consequences. The Greek philosopher Plato, writing in the fourth century BC, noted that with the removal of the trees in Attica, "there has been a constant movement of soil away from the high ground and what remains is like the skeleton of a body wasted by disease" (quoted in Thirgood, 1981).

At times, the pressure on the forests eased. The collapse of the Roman empire and the subsequent economic decline of much of Europe led to a regrowth of forests in some areas; the same happened at the time of the Black Death in the fourteenth century. But each time, as populations and economies

recovered, the forests came under attack again.

In the United Kingdom, where population growth and economic development were especially rapid, the natural forests were heavily depleted by the late Middle Ages. Iron-making, which at the time was a small-scale industry carried out in the forests, was particularly destructive. By the sixteenth century most of the accessible wood resources had been exhausted in England and iron-makers turned to the woodlands of Scotland, Ireland and Wales. Timber also began to be imported in large quantities for every purpose. The Great Fire of London in 1666 was said to have been welcomed in Bergen, Norway because of the boost the rebuilding gave to timber exports.

Laws were passed in several European countries in an effort to protect and expand wood resources, especially oak for shipbuilding and pine for masts. The laws were only partially successful, even for these limited objectives. The forest cover of France had shrunk from perhaps 35 percent of the country's area at the beginning of the sixteenth century to about 25 percent by the middle of the seventeenth century, and despite reforms of forest administration and forest laws by Louis XIV and his Minister of Finance Colbert in 1661, the situation continued to worsen because of shipbuilding and the smelting of iron ore. Nevertheless, these reforms laid the basis for a forestry tradition, a strong forest service and the many examples of woods managed sustainably for wood production that exist today.

Another illustration of the link between early forest laws and naval needs comes from Sweden, where the head of the forest service wrote in 1978 to the Swedish Navy to advise that 350 ha of oak trees planted in 1829 at the request of the Navy were now ready for collection (Swedish Forest Service, 1978). Despite these laws the forests continued to disappear under the inexorable pressures of population growth and economic development. There was also a deep feeling among many people that clearing the forests represented human progress. John Morton, an eighteenth century English writer, said: "In a country full of civilized inhabitants, timber must not be suffered to grow. It must give way to fields and pastures, which are of more immediate use and concern to life" (quoted in Thirgood, 1989).

Destruction of the natural forests took place over most of Europe. It accelerated as the demand for fuelwood and timber grew rapidly in the early years of the Industrial Revolution. But with the continuation of economic growth and industrialization, the pressures on the forests gradually eased. By the early years of the present century, the total area of the European forests had more or less stabilized, and in the past decade it has even increased slightly. In particular, the pressure on the Mediterranean forests for the supply of fuelwood and for land for agriculture has eased, although the risk of damage from fires continues.

In the United States, a similar process took place. Huge areas of forest were cleared in the eighteenth and nineteenth centuries as the population grew and industrialization got under way. Westward expansion left a trail of ghost towns with abandoned sawmills surrounded by cleared forests. But, as in Europe, the present century has seen a stabilization of the area of forest.

The reasons for this transition from rapidly diminishing to stable forest areas in the industrialized countries of the Northern Hemisphere are complex. The substitution of coal and then oil for fuelwood is one obvious factor. But more important were the major demographic and economic changes that took place as industrialization and economic growth proceeded.

Rural populations fell drastically as people flooded into the cities and towns. By the turn of the present

century, the average proportion of the population living in the rural areas of the industrialized countries was around 40 percent compared with 80 percent or more at the beginning of the Industrial Revolution. Today, in Europe and the United States the rural population is about 25 percent of the total. National population growth rates have also dropped to around replacement level in most of the industrialized countries.

Equally important were the changes that took place in agricultural production. Subsistence farming gave way to an increasingly capital-intensive form of agriculture relying on mechanization and high inputs of fertilizers and pesticides. Rather than requiring ever-expanding amounts of land, agriculture became more productive, which led eventually to the present politically charged problems of crop surpluses and the need to "set aside" or take out of

food production increasing amounts of farming land.

Although history does not repeat itself exactly, the experience of the industrialized world helps shed light on what is happening with regard to the pressures on forests in many developing countries. Expanding populations, especially when they are at subsistence level and there is significant rural–urban migration, need increasing amounts of land to feed the urban people. Even highly punitive forest-protection legislation is likely to be largely ineffectual in the face of the desperation of people trying to grow enough food to stay alive.

It is only with increased agricultural productivity and stabilized populations that the pressure to clear forest lands diminishes. This transition can be seen taking place today in some of the rapidly industrializing countries of the developing world. The Republic of Korea, for example, where the rural population has fallen sharply while agricultural production has increased dramatically, has reached the turning point at which the area of forest can begin to expand again. In other developing countries, such as Thailand and Malaysia, forest depletion continues but the economic and demographic conditions for stabilization of the remaining areas of forest are beginning to emerge.

In many other developing countries, however, rural population growth continues virtually unabated, agricultural productivity is low and stagnating economies offer few employment opportunities for those wishing to leave the land. In such cases, increasing pressure to clear forest for agriculture is inevitable.

The parallels with the experience of the industrialized world are thus obvious, but what makes today's situation in the developing world different, in a quite fundamental way, is the sheer impact of population numbers and the pace of change. For example, the population of the European continent grew from about 140 million in 1750 to 265 million by 1850, to 392 million by 1950 and to 499 million by 1990. The population of the United States grew from 23 million in 1850 to 151 million in 1950 and to 248 million in 1990 (Hoffman, 1992). Today, on the other hand, around 50 million people are being added to the rural populations of the developing world every year. At the same time, the rate at which the forests are disappearing is far faster than at any other time in human history.

Sustainable forest management will have a vital role in human welfare and development in the coming decades. But it is only a part of the

Forests are disappearing now at a faster rate than at any other time in human history; their decline is linked to rapid population growth, especially in developing countries.

response required if the rate of
depletion of the world's forests is to
be slowed and eventually halted.
What is ultimately needed is the same
transformation of economic and
demographic conditions in the
developing world that has made
stabilization of the forest area possible
in the industrialized countries.

[1]According to the FAO definition: "Forest
management deals with the overall
administrative, economic, legal, social, technical
and scientific aspects related to natural and
planted forests. It implies various degrees of
deliberate human intervention, ranging from
action aimed at safeguarding and maintaining the
forest ecosystem and its functions, to favouring
given socially or economically valuable species or
groups of species for the improved production of
goods and environmental services. Sustainable
forest management will ensure that the values
derived from the forest meet present-day needs
while at the same time ensuring their continued
availability and contribution to long-term
development needs."

CHAPTER 2

The world's forests

The world's forest resources are still substantial.
In 1980, their total area was about 3 600 million hectares.
In addition, there were 1 700 million hectares of wooded lands not classified
as forests. In all, about 40 percent of the world's land surface
is still under some type of tree cover.

But despite the immense size of this resource, there is ample reason
for concern. The estimated rate of forest depletion in the tropical zone in the
decade 1981 to 1990 was 15.4 million hectares per year, significantly
greater than the annual depletion of 11.4 million hectares assessed in the
decade 1971 to 1980. In addition, there has been severe forest
degradation over a very large area, perhaps even larger than the
area of forest depletion.

There are, however, major differences in the rates of forest
depletion in different parts of the world. The area of the temperate
and boreal forests has now broadly stabilized. It is the tropical forests
that are being so rapidly destroyed.

Temperate and boreal[1] forests

The temperate and boreal forests together make up almost half the total area of forest in the world. Virtually all the wood harvested from these forests is used for industrial purposes, with only a small proportion used for fuel. In all, the temperate and boreal forests provide over 80 percent of the world's industrial wood supplies.

The total area of temperate and subtemperate forest, including plantations, is about 760 million hectares (Lanly and Allan, 1991). The proportion of the land area occupied by forests (mainly temperate forest types) in the industrialized countries of the Northern Hemisphere is surprisingly high. In the former USSR, it is over 40 percent; in France, Germany, Italy and Poland it is 25 to 30 percent each; and in Greece and Austria it is about 45 percent. The United States has about 300 million hectares of forest, covering about one-third of its land area; the area of forest, in fact, is 50 percent greater than that of cropland.

The total area of temperate forest is now broadly stable. In the United States, for example, the decline in the total forest area between 1952 and 1987 was about 5 percent, mostly as a result of conversion of forest land for residential building, but during the same period the total standing volume of wood increased by about 25 percent. In Europe there has been a slight increase in forest area, while the total standing volume has significantly increased.

It is estimated that most of the temperate forests are under some form of management, although there is no reliable figure for the actual area. Usually the main emphasis is on producing a sustained timber yield, and foresters have a large body of research findings and practical experience dating back over 100 years on which to draw. Multipurpose management also has a long history in some of the state and communal forests in mountainous

A ponderosa pine forest in the western United States; a surprising one-third of the land area of this vast country is covered by forest.

Forest cover in the Mediterranean region has been vastly reduced by overgrazing and clearing for agriculture.

areas of western and central Europe, where it was introduced a century ago with the triple objectives of sustained timber production, soil and water conservation and the collection of fuelwood and non-wood forest products. In recent years, the conservation of ecological diversity and provision for recreational uses have become increasingly important objects of management in the temperate forests.

The Mediterranean forests form a special case within the temperate zone. They comprise, rather than a single forest type, a series of ecosystems ranging from high closed coniferous forests through all the intermediate stages to sparse arid–area scrub. Although they have been greatly reduced in area and in many areas are heavily degraded, they are generally much more complex and rich in species than the other temperate forests.

The Mediterranean region, with a population of around 1 000 million, was once heavily forested but now the forest cover is less than 5 percent of the area, concentrated mainly in the northern countries of the region. Even the 5 percent overstates the area under forest; in Morocco, for example, more than half of the 8 million hectares classified as forest are, in fact, grasslands. Clearing for agriculture and overgrazing are the main destructive agents; frequent forest fires and the harsh climate with its periodic droughts are additional pressures on the forest resources of this region.

While the area of the temperate forest types may be stable or even increasing, there is a loss of quality in some places. "Quality" refers to such attributes as health, environmental and social benefits and comparability of substitutes such as plantations with natural associations (see e.g. Dudley, 1992). Dieback and decline[2] over large areas of temperate forests have been seen as symptoms of poor health, caused by a wide variety of factors including climatic stress and air pollution. Dieback and decline can also occur in tropical and subtropical forests.

The boreal forests encircle the northern part of the globe, extending through Alaska (USA) and the northern regions of Canada, Finland, Norway, Sweden and the former USSR. Their area is vast, a total of about 920 million hectares of closed forest with a further 300 million hectares of open woodland. Almost three-quarters of the total is in the former USSR. A further 20 percent is in Canada and Alaska and about 5 percent in the Nordic countries.

Production from the boreal forests accounts for about 50 percent of the world's newsprint, nearly 40 percent of the sawnwood and 20 percent of the paper pulp. Although large areas of forest are harvested every year, the quantities of wood removed are substantially less than the overall annual growth of the boreal forests. Because of the harsh climate, difficult and often dangerous terrain and long distances, a large proportion of these forests is likely to remain outside the scope of commercial logging for the indefinite future.

Since the early part of the present century, the boreal forests have been actively managed for sustained yield in the Nordic countries. In Canada and the former USSR, in contrast, control over logging has generally been weak, although significant efforts to improve the position have been made in recent years. The political and economic uncertainties in the former USSR, and the manner in which these are likely to be reflected in the management of the region's huge areas of boreal forest, are at present a major cause of concern.

In the boreal forests, much of the annual precipitation is in the form of snow.

Tropical forests

The tropics are, geographically speaking, the area between the Tropic of Cancer and the Tropic of Capricorn, 23° 30' south and north of the equator, respectively. The area includes most of Africa; Mexico, Central America and South America to about the latitude of Rio de Janeiro (Brazil); most of the Indian subcontinent, southern China, Southeast Asia and the northern half of Australia. Within this vast area there is a wide range of forests, but they can be broadly divided into four main types: the lowland formations, comprising the tropical rain forests; the moist deciduous forests; the dry and very dry forest zones; and the upland formations. In total, the tropical forests cover about 1 700 million hectares, an area roughly that of South America.

Only a small proportion of the world's tropical forests are under management in any meaningful sense.[3] Even where there is management, it is often confined to the collection of revenue from logging operations or the protection of national parks by government forest services. Where these forest services are weak and ill-equipped and the staff is poorly paid the beneficial impact of such "management" is likely to be small.

As public awareness of tropical

Africa has a diverse range of forests and woodlands; trees in a typical landscape on the outskirts of Karuzi, Burundi.

deforestation has grown, the main public focus has, misleadingly, been on forest loss caused by logging. In fact, the main reason for tropical deforestation is agricultural expansion, which in all its forms, from shifting cultivation to cattle ranching, accounts for about 90 percent of the total forest loss presently taking place. Furthermore, and contrary to popular perception, the largest losses of forest area are not occurring in the tropical moist forests but in the tropical moist deciduous forests, while the most rapid rates of deforestation are occurring in the moist deciduous forests, the dry zone and the upland forests, as the following sections will show.

In the decade between 1981 and 1990 there was an average annual loss of forest in the tropical zone of 15.4

million hectares yearly, or an annual rate of loss of 0.8 percent. The situation is summarized in Table 1.

Tropical rain forests

The tropical rain forests are so called because they occur in areas with annual rainfall of more than about 2 500 mm. They are evergreen, luxuriant and rich in tree species as well as in other plant and animal life. They are a major source of the world's hardwoods; mahogany, rosewood, ebony and sapele are the most popularly known but there are dozens of other species which are used in fine furniture, veneers and other high-grade uses. There are also many other species of lesser commercial value which are used for construction, plywood and other purposes.

The soils on which the rain forests grow tend to be shallow and poor in nutrients. The high growth rates and intense biological activity of the forest is based upon a quick recycling of organic waste material back into the growing plants. In many areas, if the trees are removed and the land is used for agriculture, the nutrient cycle is broken and the inherent lack of fertility of the soil is quickly revealed. Crop yields fall after a few years and the land may become susceptible to erosion or laterization – the formation of a hard infertile crust.

Table 2 shows that the total land area of the rain forest zone is 947 million hectares, of which 718 million are covered by forest. The moist forests of Africa remain largely intact; they still occupy 73 percent of the zone. Contrary to what is widely believed, the same is broadly true in Latin America, where 87 percent of the zone is forested. In Asia, however, tropical moist forest depletion has proceeded further, and just over half of the land area of the zone is forested.

By far the greatest concentration of the tropical rain forest is in the Amazon Basin. This huge area of forest, which covers northwestern Brazil and stretches into the neighbouring countries of Colombia, Ecuador, Peru and Venezuela, accounts for two-thirds of the world's tropical moist forest. Asia has the next largest area, mostly in Indonesia, Malaysia and Papua New Guinea. Africa has about 15 percent of the total, almost entirely in the Congo and Zaire.

The total area of rain forest cut from 1981 to 1990 was 4.6 million hectares per year or, more graphically, nearly 9 ha per minute. This gives an overall rate of deforestation of 0.6 percent per year, but the rate was almost twice this in Asia.

Mangrove forests are included in the area of tropical rain forest; it is

Nurturing seedlings in Latin America – land clearing for agriculture breaks the nutrient cycle of the soil and may lead to erosion.

estimated that there are about 24 million hectares of coastal mangrove forests in the subtropical and tropical countries (FAO, 1988a). Human pressure to overcut mangrove forests for the provision of wood products or to clear them for salt pans and for infrastructure development is heavy, particularly near urban areas, despite the many products and important protective functions that they provide.

Moist deciduous forests

The moist deciduous forests occur in areas with annual rainfall of 1 000 to 2 000 mm. These forests vary greatly depending on factors such as the

amount of rain and its distribution throughout the year, the temperature and the type of soil. They are generally much less rich in tree species and less biologically diverse than the tropical moist forests.

The total area of the zone is nearly 1 300 million hectares. Because the zone is well suited for human settlement, deforestation has proceeded much further than in the rain forest zone. The proportion of the area still forested is 46 percent, although it is only 29 percent in Asia (see Table 3). The actual forest area is less than that of the rain forests, with most of it shared between Africa and Latin America.

The rate of loss of these forests in the decade 1981 to 1990 has been enormous, at about 6.1 million hectares per year.

Forests of the dry and very dry zones

The tropical dry forests occur in areas with rainfall of 500 to 1 000 mm per year. They are relatively open and, especially in Africa, are rich in wildlife. The people living in and around these forests tend to rely heavily on them for livestock grazing as well as for fuelwood, building poles and other products. Many of these forests are fragile, and improper exploitation, even if light, can lead to severe degradation, weed infestation and increased susceptibility to fire and insect damage. Large areas have degenerated into grass and scrubland.

More than half the world's dry tropical forests and savannah woodlands are in Africa (see Table 4). Its distinctive savannah landscape stretches all across the continent to the south of the Sahel and covers much of eastern and southern central Africa. Dry forest zones are found also in northeastern Brazil and much of India. In many areas, these forests merge into arid or even desert margin zones where the natural tree cover becomes increasingly sparse.

The total area occupied by the dry forest zone is roughly one-third that of the rain forest zone but depletion has reached an advanced stage, especially in Asia where only 15 percent of the zone is now forested.

Tropical upland formations

Forests occurring at an altitude of 800 m or above are referred to as upland or sometimes hill and montane forests. Their characteristics can vary greatly depending on their altitude, rainfall, temperature and other factors.

The upland forest zone is surprisingly extensive (see Table 5). Its area is almost as great as that of the forests of the dry and very dry zones. In Asia the forests of the zone cover the Himalayas and stretch south into Myanmar and the mountains of

Thailand and Viet Nam. In Latin America, the zone includes the Andes and the highlands of Mexico, and in Africa it covers the highlands of Ethiopia and the mountainous areas around Lake Victoria.

Depletion of these forests has reached an advanced stage, especially in Asia and Africa, and is continuing at a rapid rate.

Deforestation and its impacts

The term deforestation summons images of devastated landscapes, eroded soils, desertification and human misery. Yet visitors to highly deforested areas, for example in parts of the Sahel,[4] are surprised to see landscapes in which many trees still remain. Such confusion arises from the way in which the term deforestation is used.

A forest in the tropics, according to the FAO definition,[5] is an area of land of which at least 10 percent is covered by the crowns of the trees or bamboos growing upon it and which is not subject to agricultural practice. Deforestation implies a reduction in the crown cover to less than 10 percent or a change in land use. Thus an area of well-wooded productive agricultural land will be officially classified as deforested.

An area of forest that has been badly damaged by, for example, careless logging but that has not been converted to another use and still retains at least 10 percent crown cover, is designated as degraded. It is only if it is taken over for agriculture or other uses or if the crown cover falls below 10 percent that it is described as deforested. Outside these FAO definitions, the terms deforestation and forest degradation are used much more loosely and often interchangeably.

All interference with a natural forest alters its ecology to some extent, and many of the world's forests have been subject to a considerable amount of human interference. In Europe, for example, virtually none of the original forest survives. Traditional forest-dwelling peoples, though they may leave much of the structure of the forest intact, can also bring major changes in the species composition as well as the wildlife through selective utilization. The bulk of the world's remaining untouched forests are in the far north of Canada and the former USSR, areas too remote and inhospitable for human habitation, or

in the dense moist forests of the tropics as in the Amazon and Zaire river basins.

Logging, which is often popularly equated with deforestation, does not normally bring about the disappearance of forests. Logging in the tropics is selective, removing relatively few stems per hectare, in contrast to logging of natural forest in, for example, the United States or Canada, where it is closer to clear felling. But removal of even some of the trees, no matter how carefully done, obviously alters the spatial and size class distribution of tree species in a forest, although this is a long way from eliminating the forest completely. Provided the area is not converted to other uses and the soil cover remains intact, new trees replace those that have been felled and in a few decades the difference is only discernible to the expert eye or revealed by forest inventory data.

Improper logging can, however, seriously degrade a forest. This is particularly true if the logging is carelessly carried out on steeply sloping ground or if the access roads have been poorly built or positioned so that heavy erosion occurs. But trees can usually manage to establish themselves again, even in badly eroded areas within the forest, provided they are protected; this leads eventually to the restoration of forest cover.

Although logging is often popularly equated with deforestation, it is a misconception that it directly causes the disappearance of forests.

A decade ago, cutting for fuelwood was widely believed to be one of the major causes of deforestation. This is not generally the case, although it is true that heavy cutting for charcoal-making or urban fuelwood supplies can result in forest clearance, especially in savannah woodland. In the rural areas, rather than felling trees for fuelwood, most families tend to harvest branches of trees or shrubs on a sustainable basis. In comparison with clearing for agriculture, the impact of rural fuelwood collection is usually small.

But even clearing for agriculture does not necessarily lead to the complete disappearance of forests. Many traditional farming systems incorporate a fallow system. An area of forest is cut and burned, usually leaving some trees still standing. It is farmed for a few years and then left fallow while the family moves on to another location. Eventually, the tree cover regrows on the previously farmed area. Shifting cultivation can be a stable and effective land-management system which preserves much of the tree cover as well as the fertility of the land, provided the fallow period is long enough.

It is when population density begins to increase that problems arise with fallow systems. Because there is no longer enough land for everyone, farmers find they are unable to allow fallow areas sufficient time to restore the fertility of the land fully. The soils are progressively degraded and in some cases are colonized by weeds. This has happened, for example, in Asia, where tens of millions of hectares have been lost to cogon grass *(Imperata cylindrica)*.

Similar problems occur when forest land that is not suitable for agriculture is cleared. If the cleared land is used for subsistence farming, in which there are no fertilizer inputs, then the fertility of the soil is quickly exhausted and its structure is destroyed. The end result is an area of degraded land capable only of supporting a sparse growth of scrub or weeds or, where this is lacking, often subject to heavy wind or water erosion.

On the other hand, much of the forest clearing carried out in the past has led to the creation of fertile and sustainable areas of agricultural and grazing land, as has been shown in virtually every country in the world. The human race depends for the bulk of its food supplies on such areas. When, therefore, the conversion of forest areas that have the potential to support permanent agriculture is properly planned and the subsequent cultivation methods are appropriate, the transition can be both successful and highly beneficial.

[1] The word comes from classical Greek and refers to the north wind. It is applied to the forests of the northern part of the globe.

[2] Dieback and decline are terms used interchangeably to describe an episodic event characterized by premature loss of tree and stand vigour and health, without any obvious sign of physical disturbance, disease or insect attack. Symptoms are progressive and non-specific. They may include growth reduction, rootlet mortality, abnormally small or discoloured foliage, premature senescence of foliage, dead branches, thin crowns and epicormic branches. Many causal factors, acting alone or in combination, can cause dieback or decline.

Both the terms dieback and decline have been used to describe disease symptoms, with dieback usually referring to the death of branches and decline referring to a more general set of symptoms associated with loss of tree vigour (FAO, 1993f).

[3] The actual area under management is, just as in the temperate zone forests, not exactly known. A much-quoted study carried out for the International Tropical Timber Organization in 1988 found that the area of natural moist forest subject to sustained yield management in the organization's tropical Southeast Asian member countries was 1 million hectares of a total of 828 million hectares, or just over 0.1 percent. The area of managed forest is, however, disputed by several countries which state that their forests are under sustained yield management. Whatever the true situation, the proportion of forests under any form of management in the tropics (and subtropics) is likely to be less than in the temperate zones, and the argument over the actual proportion of forests that is managed in the tropical, subtropical, temperate or boreal zones reveals the acute shortage of reliable information.

[4] Examples can be seen in Senegal, where *Faidherbia albida* dominates the landscape described as *paysages à parc* (parkland) and *Borassus flabellifer* dominates the *vignoble serere*. Such parkland with *F. albida* is seen also in Mali and Burkina Faso (often with *Butyrospermum paradoxum* and *Parkia biglobosa*) as well as in Niger and northern Cameroon. Unfortunately there are now few examples where the trees are being actively regenerated by the farmers.

[5] "In the case of industrialized countries the word 'forest' is defined as 'land with tree crown cover of more than about 20 percent of the area. Continuous forest with trees usually growing to more than about 7 m in height and able to produce wood. This includes both closed forest formations where trees of various storeys and undergrowth cover a high proportion of the ground and open forest formations with a continuous grass layer in which tree synusia cover at least 10 percent of the ground'. In the case of tropical countries 'forests' are defined as 'ecological systems with a minimum of 10 percent crown coverage of trees and/or bamboos, generally associated with wild flora, fauna and natural soil conditions, and not subject to agricultural practices'." (FAO, 1992b).

Table 1 **Forest cover and deforestation in the tropical zone**

Region	Number of countries	Land area (million ha)	Forest area 1980 (million ha)	Forest area 1990 (million ha)	Annual change 1981–90 (million ha)	Annual rate of change (%)
Africa	40	2 236	568	527	−4.1	−0.7
Latin America	33	1 650	992	918	−7.4	−0.8
Asia	17	892	350	311	−3.9	−1.2
World total	90	4 778	1 910	1 756	−15.4	−0.8

Note: Table 1 gives figures on forest cover in the tropical zone as a whole, including forests growing in zones not regarded as zones of natural forest growth, such as deserts or alpine areas. The data for Tables 2 to 5 are restricted to the zones of natural forest growth. The sums of the figures in these tables do not necessarily agree with Table 1.
Source: FAO, 1993c.

Table 2 **Deforestation in the tropical rain forest zone**

Region	Total land area of zone (million ha)	Total forested area 1990 (million ha)	Total forested area 1990 (% of zone)	Annual deforestation 1981–90 (million ha)	Annual deforestation 1981–90 (% of zone)
Africa	118.5	86.6	73	0.5	0.5
Asia	306.0	177.4	58	2.2	1.1
Latin America	522.6	454.3	87	1.9	0.4
World total	947.1	718.3	76	4.6	0.6

Source: FAO, 1993c.

Table 3 **Deforestation in the moist deciduous forest zone**

Region	Total land area of zone (million ha)	Total forested area 1990 (million ha)	Total forested area 1990 (% of zone)	Annual deforestation 1981–90 (million ha)	Annual deforestation 1981–90 (% of zone)
Africa	653.6	251.1	38	2.2	0.9
Asia	144.6	41.8	29	0.7	1.5
Latin America	491.0	294.3	60	3.2	1.0
World total	1 289.2	587.2	46	6.1	1.0

Source: FAO, 1993c.

Table 4 **Deforestation in the dry and very dry zones**

Region	Total land area of zone (million ha)	Total forested area 1990		Annual deforestation 1981–90	
		(million ha)	(% of zone)	(million ha)	(% of zone)
Africa	823.1	151.2	18	1.1	0.7
Asia	280.6	41.1	15	0.5	1.1
Latin America	145.4	46.0	32	0.6	1.3
World total	1 249.1	238.3	19	2.2	0.9

Source: FAO, 1993c.

Table 5 **Deforestation in the tropical upland formations**

Region	Total land area of zone (million ha)	Total forested area 1990		Annual deforestation 1981–90	
		(million ha)	(% of zone)	(million ha)	(% of zone)
Africa	169.2	35.3	21	0.3	0.8
Asia	102.6	47.2	46	0.6	1.2
Latin America	429.1	121.9	28	1.6	1.2
World total	700.9	204.4	29	2.5	1.1

Source: FAO, 1993c.

CHAPTER 3

Why are forests important?

Forests provide a wide range of benefits at local,
national and global levels. Some of these benefits depend on
the forest being left untouched or subject to minimal interference.
Others can only be realized by harvesting the forest for
wood and other products. Yet other benefits from forests,
despite being frequently claimed, are illusory.

Forests for wood and energy

Wood is one of the most versatile and ubiquitous products in human use. Wood and the products derived from it are found in every area of modern existence, from the timber used in construction, furniture and a myriad of industrial and domestic uses to fibre board, chipboard, paper, newsprint and cardboard.

As a construction material, wood is strong, light, durable, flexible and easily worked. It has excellent insulating properties. In contrast to the substitutes for wood in structural and architectural uses such as brick, concrete, metals and plastics, wood can be produced and transported with little energy consumed and the products are renewable and usually biodegradable (Koch, 1991).

Wood is also of major economic importance. The world demand for timber and wood products has been growing at 1 to 2 percent per year over the past decade. The total world production of industrial timber in 1990 was about 1 600 million cubic metres, of which about three-quarters came from the industrialized countries. Paper production was about 235 million tonnes and wood pulp production was 160 million tonnes; in both cases, over 80 percent of the production was in the industrialized world.

The present international trade in roundwood, sawnwood and wood products is worth about US$36 000 million per year, of which about US$10 000 million comes from developing country exports. The trade in paper and pulp accounts for a further US$60 000 million, of which the share of the developing world is just US$2 500 million. The overwhelming majority of the world's exports of wood and wood products comes from the industrialized rather than the developing countries, contrary to popular belief.

The 1990 estimate by FAO of world fuelwood and charcoal consumption was about 1 800 million cubic metres (FAO, 1992a). Of this, nearly 90 percent was in the developing world, where for many people wood meets virtually all energy needs. Since most of the fuelwood consumption and charcoal-making occur on a small-scale, informal basis in rural areas, where no statistics are kept, any such estimates must be treated with considerable caution. The figures nevertheless suggest that the world consumption of wood for energy is roughly similar to that of industrial wood.

Non-wood forest products

In traditional forest management, forests have primarily been seen as wood-producing units and other products have conventionally been referred to as "minor forest products"; their supply was not given high priority. It is only in recent years that proper attention has been paid to the importance of these products, especially in their local context where they may be considerably more valuable than the wood obtained from the forest. It is now becoming accepted that in many cases recognition of the economic and social importance of these other forest products may be the key to the active involvement of people in forest management.

Non-wood forest products include plants for food and medicinal purposes, fibres, dyes, animal fodder and other necessities. For instance, the Kayapó people of Gorotire village in southern Pará State, Brazil, utilize over 98 percent of the 120 species occurring

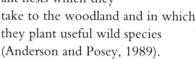

within the local scrub savannah (campo cerrado). The Kayapó even prepare a planting medium from litter and termite and ant nests which they take to the woodland and in which they plant useful wild species (Anderson and Posey, 1989).

The wildlife in forests can also make a major contribution to food supplies. Surveys in Cameroon, Côte d'Ivoire, Ghana and Liberia found that forest wildlife accounted for 70 to 90 percent of the total animal protein consumed. People surveyed in one area in Ghana considered the loss of such meat supplies to be the worst impact of local forest destruction (FAO, 1993g).

Forest products may also be commercially important at a local level, where they are traded in markets and shops or are sent to the larger towns and cities. In India, for example, the tendu leaf *(Diospyros melanoxylon)* is used for the locally made bidi cigarette. It is estimated that the collection and processing of the leaves provides part-time employment for up to half a million women.

Some of these forest products are valuable export commodities. They include gums and resins, bamboos, various oils, turpentine, tanning

Non-wood forest products – including food and fodder – are highly valuable, especially in local areas; recognizing their importance may encourage people to become actively involved in forest management.

materials, honey, spices, bark and leaves and medicinal plants. Rattan, the long thin stem of a climbing palm (mainly *Calamus* sp.), has become an important export for Indonesia, Malaysia and the Philippines. Portugal, Morocco and other Mediterranean countries export large quantities of cork derived from the cork oak, *Quercus suber*. The Republic of Korea has built up an export trade in edible forest fungi, while gum arabic from *Acacia senegal* has long been an important export from Sudan.

Forests as dwelling places

Forests and woodlands provide a dwelling place for more than 200 million people in the tropics (Brown *et al.*, 1991). They include those who have lived there for generations, often referred to as indigenous or tribal peoples; people who have recently moved into the area, often described as settlers, squatters or encroachers; and people who live part time in the forest working as small-scale loggers or harvesters of forest products. The numbers vary with time and among different areas, but all need to be taken into account when forest management is being considered.

Indigenous people, contrary to widespread assumption, are not only hunter-gatherers with minimal impact on the forests, but may be shifting agriculturalists as well. Their traditional slash-and-burn fallow systems have generally provided them with a stable existence, though at a low standard of living, while at the same time retaining the basic forest structure. But the decreasing area of forest available for farming and rising populations are resulting in shorter fallow periods, and the system is becoming less effective in maintaining the people.

Settlers are frequently "shifted" agriculturalists who have been unable to find the land required to grow enough food in the area in which they originate. When they move into a forest, because they lack the local knowledge and traditional skills of indigenous peoples, they often tend to be more destructive, quickly exhausting a patch of land through cultivation techniques that are inappropriate to the local conditions and then moving on to repeat the process elsewhere. Some may be interested mainly in clearing an area and selling the timber in order to raise enough capital to start a small business in the city. Others may be

sponsored by wealthier people wishing to extend their landholdings by clearing and laying claim to areas of forest.

Because they lack heavy equipment, most of these settlers are unable to penetrate deeply into dense forests; they therefore tend to operate around the edges. But when roads are built into the forest for logging, mining or general transport they provide settlers with easy access. Much of the forest encroachment taking place in the humid tropics is along such routes.

Environmental benefits of forests

Forests and woodlands have an important role in protecting the environment at a local and even regional level. This is particularly true of steeply sloping watersheds where the tree roots are important in binding the soil and protecting it against erosion and landslide.

Uncontrolled clearing of forests from such upland areas, in addition to its local effects, can also have major repercussions further downstream. The eroded soil carried by streams and rivers is mainly deposited in reservoirs for irrigation and hydroelectricity and reduces the capacity and shortens the life of these costly investments.

Forests can also play a major part in areas that are covered with snow in

Forests are havens for wildlife such as this vervet monkey, which inhabits the forests of Kenya.

winter. During the spring, forests help regulate the rate at which the snow melts and also reduce the danger of avalanches. In Colorado, USA, for example, the regulation of the snow melting in the spring is seen as the most important benefit from the forest cover on slopes.

At a local level trees can also bring many environmental benefits. They provide protection against wind erosion. They can help increase the rate at which rainwater infiltrates and recharges the groundwater. Used judiciously in farming systems they help maintain the fertility of the soil as the nutrients drawn up by their roots are recycled into the top layers of the soil by leaf fall. They provide shade for animals and humans; the microclimate under trees may be several degrees cooler and more humid than out of their shade. In coastal areas, mangroves protect the land against erosion by the sea as well as providing breeding grounds for fish and shrimps.

Forests also have an increasingly important role as havens for wildlife and for the protection of endangered species of plants and animals. Often allied to this is their role in recreation, tourism and what has come to be known as "ecotourism". Hiking, camping, nature study and simply getting out of the city are increasingly important to urban people in their often stressed and polluted world. Aesthetic appreciation of trees and forests as well as the cultural and spiritual values that they epitomize are very important to rural and urban dwellers alike.

The development of ecotourism may pay significant dividends. A wide public has been sensitized to nature and the environment by books and television documentaries. Many want to see for themselves and are willing to pay substantially to do so. Game-viewing safaris are already well established and cater to a wide market. More specialized holidays, offering hiking, camping, bird-watching and opportunities to study nature in detail, are becoming steadily more popular. To tap this market by providing an increasingly knowledgeable public with what it wants, while at the same time avoiding creating social or environmental problems, requires a continued and expanding development effort. Measures have to be taken to ensure that the money generated by ecotourism benefits the country and the communities concerned and is invested in conservation of the resource itself.

The red deer is one of the several wildlife species associated with the forests of Europe.

Genetic resources and biodiversity

The term "genetic resources" refers to the economic, scientific or social value of the genetic variation found among and between species. If properly managed, these resources are renewable; they can be used, without ever being used up (Ledig, 1986).

Genetic variation has a number of fundamentally important functions; it constitutes a buffer against changes in the environment (including those brought about by pests, diseases and climate change) and it provides humans with the building blocks for selection and breeding to adapt plants and animals to a range of environments and end uses (Palmberg, 1987). Intensive selection and breeding for increased yields and uniformity has long been practised for agricultural crop species. While such selection can improve certain traits in the short term, it may also reduce genetic variation when done over many plant generations, and the more genetically uniform populations have a decreased ability to respond to changing environmental pressures (including attack by pests and diseases) or shifting human needs. Therefore the use of narrowly based genetic materials grown for short-term productive purposes must always be paralleled by conservation of the maximum amount of available variation through the establishment of reserves and managed resource areas and through the inclusion of genetic conservation concerns in improvement and breeding strategies.

An example of the adverse effects of a narrow genetic base has recently been seen in the decline of the widely planted neem tree (*Azadirachta indica*) in Sahelian countries, believed to be due to environmental stress (Ciesla, 1993). Narrowing of the genetic base occurred because the neem seed quickly loses viability, which restricted the original importations to the region from its native Asia. The effect of this may have been to reduce the tolerance of Sahelian populations of the species to moisture stress. Another example has recently been provided by the widespread planting of the giant varieties of ipil ipil (*Leucaena leucocephala*) in the tropics. The most widely used varieties were propagated from a small number of individuals. They have recently come under heavy attack by the Leucaena psyllid (*Heteropsylla cubana*), an insect which has spread rapidly from its native habitat in Mexico and Central America to the Pacific Islands, Asia and, most recently, eastern Africa.

Narrowing of the genetic base in tree breeding is carried to its extreme in clonal forestry, where there is no genetic variation among individuals in each monoclonal block. Poplars and willows have long been selected for commercial properties and desirable individuals have been propagated through cuttings, but the number of widely used clones has been relatively limited. Insects and diseases have quickly adapted, with the result that most of the early clones have been heavily attacked. Poplar breeders now give first priority in their breeding strategy to resistance to such attacks, and only secondarily consider attributes such as growth rate, form or wood properties. This strategy relies entirely on the ability to return repeatedly to the gene pool of the wild species in the search for genes conferring resistance.

The world's forests, especially those in the tropics, are both laboratories for the natural selection of genetic resources of plants and animals, on a scale which cannot be matched by today's or any conceivable future research stations, and dynamic storage banks for those genes. In an era of increasing pressure on resources and significantly changing environmental conditions, they provide one of humanity's most effective ways of buffering itself against a highly uncertain future.

Biodiversity (or biological diversity) refers to the variety of life forms, the ecological roles they perform and the genetic diversity they contain

(Wilcox, 1984). It is estimated that the tropical forests contain at least 50 percent and probably a considerably higher proportion of all the living species on the planet, including a great proportion of higher plants and mammals. There are, for example, 50 indigenous tree species in Europe north of the Alps. In Malaysia, in contrast, an area of forest covering just 50 ha was found to contain 830 tree species, and in Peru nearly 300 species of trees have been recorded on a single hectare (Whitmore, 1990).

No financial value can be placed directly on the almost infinite variety of living species in the forests. Only a tiny proportion are ever likely to be

Many common products originate from forests; here, resin is tapped from pine trees in Viet Nam.

studied in detail, let alone be found useful to humanity. Yet the loss of each individual species makes the world a biologically poorer place. It must also be understood that such a loss affects the interlinkages and symbiotic relationships with other species.

It is possible to point to some very tangible gains that have been obtained from the resources other than wood available in the world's forests. Many common products, rubber for example, are of forest origin; so also are various fruits. Various widely used medicines were originally discovered by analysis of forest plants, often those traditionally used by forest dwellers.

There is no reason to suppose that this flow of beneficial products from the forests will be reduced in the future, provided that due attention is given to their conservation and sustainable management. The natural ecological systems of the forests will continue to evolve as they have always done. In the continuing human battle against diseases and pests the need for new products from the forests is likely to be as great in the future as it has been until now.

Forests and global climate

There is no doubt about the reality of the greenhouse effect. Certain gases in the atmosphere trap heat which

would otherwise be radiated into space. Without this naturally occurring greenhouse effect, the average temperature of the earth would be about 30 °C lower than it is.

There is also no doubt that the proportion of greenhouse gases in the earth's atmosphere is increasing as a result of human activities. The main increase to date has been in carbon dioxide. If the increase continues at the same rate as in recent decades, the global warming that is likely to result could ultimately have severe, if not catastrophic, consequences for hundreds of millions of people.[1]

The main contribution to the increasing concentration of greenhouse gases in the atmosphere has been from the burning of fossil fuels. This is likely to remain the case over the next few decades. The Intergovernmental Panel on Climate Change (IPCC) projected that fossil fuel consumption would contribute 65 percent of the greenhouse effect from 1990 to 2025. The contribution from deforestation and biomass burning over the same period is projected to be 15 percent (IPCC, 1990).

Because trees, like all green plants, absorb carbon dioxide during photosynthesis, reforestation is often suggested as a means of countering the increase in greenhouse gases in the atmosphere.[2] But trees only absorb carbon dioxide when they are growing. Carbon uptake is greatest in the early years when the rate of growth of the tree is at its maximum and tapers off as the tree reaches maturity. Eventually the trees in the forest simply act as a carbon store when no further net growth of the forest is taking place. When the trees die or are harvested, a portion of the carbon stored as woody tissue is once again released into the atmosphere; the portion of carbon released varies with the product made from the tree. If the tree is burned as fuelwood then a high proportion of fixed carbon is released, but if it is made into a durable product with a long-term use (such as furniture) then the carbon will remain fixed for a long time.

The areas of forest plantation required to have a significant impact on the amount of carbon dioxide in the atmosphere are colossal. For example, one study (Sedjo and Solomon, 1989) indicates that the current net annual increase in atmospheric carbon (approximately 3 000 million tonnes) could be sequestered in approximately 465 million hectares of plantation forests for about 30 years, or as long as they remained alive. This corresponds to an increase of more than 10 percent in the current area of all forests on the earth's surface or an increase of more than four times the present plantation area in the world to sequester only the current net annual increase in atmospheric carbon. Even this enormous estimate is based on the assumption of an average annual growth of 15 m^3 per hectare per year, which is unlikely to be achieved in temperate regions.

The effect of the planting of trees outside closed forests on the global climate and on carbon sequestration will depend upon its scale. Large-scale tree planting projects in agricultural areas, such as the "four around" schemes in China reported to have been carried out on 6.5 million hectares of agricultural land in the decade 1981 to 1990, must sequester considerable quantities of carbon, as well as providing other environmental benefits.

Urban trees, in contrast, sequester minor amounts of carbon but can make other contributions to reducing the effects of climate change which are much more significant. A well-placed urban tree is about 15 times more valuable than a forest tree from the carbon cycle perspective. Urban trees break up "heat islands" by providing shade. This can lessen air-conditioning use which requires inputs of fossil fuels. Heating during winter months can also be reduced because trees can provide shielding from winds. It has been estimated that three strategically placed trees per

Some 15 percent of the greenhouse effect from 1990 to 2025 will be caused by deforestation and biomass burning.

house can reduce home air conditioning needs by 10 to 15 percent, especially in developed areas of subtropical and middle latitudes (FAO, 1990a).

There is great uncertainty about virtually all aspects of the global warming question. It will probably be at least ten years, and some scientists now say 30 to 40 years, before data are adequate for firm scientific conclusions on what exactly is happening and what its effects are likely to be at the regional level and below. In the meantime, the high stakes involved indicate a need for prudence on the part of humanity. Both slowing the rate of deforestation and planting trees can be important in reducing the risk of global warming, and both actions would simultaneously produce a range of other benefits for humanity.

If, at the same time, the necessary actions to curb the emission of greenhouse gases from all sources are not taken, and global warming occurs on a significant scale, there could be far-reaching consequences for the world's forests. On the one hand, an increase in atmospheric carbon and higher temperatures could result in increased growth rates. The natural ranges of many tree species could advance to higher latitudes or higher altitudes as temperatures increased. On the other hand, those forests that

were stressed by climate change would become more susceptible to damage by fire, insects, pollution and disease. Genetic variation could be greatly reduced, leaving only the most resistant genotypes, and large areas of forest and even tree species could be lost.

Forest myths

There is no doubt about the immense importance of forests and the many benefits they can bring. There is, therefore, no need for untrue or exaggerated claims. Bad arguments can undermine good cases and can make advocates for forests look less credible in the eyes of knowledgeable people.

The belief that forests increase local rainfall, for example, is almost invariably false. Most rain is carried by air currents to the area on which it falls and is usually derived from ocean evaporation; droughts are caused by changes in the patterns of the rain-bearing winds. Denuding an area of trees will have little or no effect on whether rain-bearing winds blow across it or rain falls upon it.

The belief that forests provide additional oxygen supplies to the areas where they are growing – sometimes suggested as one of the advantages of peri-urban plantations – is totally false. During the process of photosynthesis, oxygen is indeed emitted. But relative to the volume of the atmosphere, the amounts are tiny and the increase in the oxygen content of the air in a forest is negligible. Moreover, during the hours of darkness, trees are net absorbers of oxygen and emitters of carbon dioxide.

The notion that forests regulate the flow of streams and rivers by acting as sponges, absorbing rainwater and releasing it gradually, sometimes believed to be a seasonal phenomenon, is also false. In the case of light rainfall, a forest intercepts a high proportion of the water, preventing it from reaching the ground. With heavier rainfall, the forest does initially slow down the rate of water runoff and this will tend to increase infiltration, though this will be at least partly balanced by the water taken up by the roots and emitted through the leaves. In very heavy rain, the ground beneath the tree is quickly soaked with water. Once this has happened, any additional water simply flows away into the streams and watercourses. The severity of floods that occur in the lower areas of river systems is primarily determined by the amount of rainfall and how it is distributed over the river basin. A distribution that results in all the tributaries or rivers flowing into the flood plain at the same time will produce the worst flooding.

The belief that trees are always benign elements in the farming landscape is also a myth. Trees can compete with crops for light and nutrients, causing significant reductions in crop yields. They can spread as weeds, ruining the land for grazing or crops. They can lower the water table, causing wells and water-holes to dry up. Poorly designed windbreaks can cause funnelling of winds, increasing plant damage and soil erosion. Equally, ill-chosen or badly positioned trees on hillsides can increase water erosion. They can harbour tsetse fly and highly destructive pests such as the *Quelea* bird.

A dramatic and tragic example of the damage that can be caused by poorly planned tree planting is the November 1988 flooding in Thailand. Large numbers of people were killed as whole villages were buried in mud and tree trunks swept down from the hills above them. The blame was placed upon careless logging, and so great was the public outrage that further logging was banned in the country's primary forests. It is now emerging that the real cause of the disaster was the expansion of rubber plantations on steep slopes. Cloned rubber saplings were planted on terraces constructed not for erosion control but for ease of rubber tapping. When the floods came, the terraces and the trees growing on them were swept down on to the villages below.

Care is always needed in the selection and planting of trees, not only to match the species to the site and to the uses to which people expect to put them, but also to anticipate their effects, especially on agricultural land. The objections or lack of enthusiasm shown by local people for many of the initiatives so eagerly pressed upon them by tree planting enthusiasts are not always as ill-founded as has tended to be assumed in the past.

[1]"...it is reasonable to conclude... that human activities have changed the composition of the atmosphere and these changes, by exceeding the buffering capacity of the earth's ecosystems, will have a major impact on the forest ecosystem. Forest management practices need to adapt to this change.

"A number of Global Circulation Models (GCMs) have been developed to better understand the changing climate, especially with elevated levels of CO_2. Increases of global mean temperature in the range of 1.5–4.5 °C are based on a doubling of carbon dioxide levels and exceed previous historical rates. Global precipitation is expected to increase slightly but the net effect will be a significant increase in evaporation rates in the middle latitudes of the northern hemisphere. ...all GCMs agree that the future climate direction is towards global warming. This conclusion alone is sufficient to shape future forest management actions" (FAO, 1993h).

[2]The rate of carbon fixing is a function of many variables. These include tree species, growth rates, longevity, site, annual precipitation and length of growing season. Annual rate of carbon fixing is highest in young plantations. Fixation rates for several tropical forest plantation species over a given rotation are summarized by Schroeder (1991) as follows:

Species	Rotation (years)	Mean carbon storage (t/ha)
Pinus caribaea	15	59
Leucaena sp.	7–8	21–42
Casuarina sp.	10	21–55
Pinus patula	20	72
Cupressus lusitanica	20	57
Acacia nilotica	10–15	12–17

Data for temperate zone species indicates that plantations of *Acer saccharinum* can absorb 5 tonnes C per year under optimal conditions and that *Platanus occidentalis* absorbs as much 7.5 tonnes C per hectare per year (FAO, 1990a). Carbon is also stored to varying degrees in forest soils.

CHAPTER 4

Dealing with conflicting interests

The world's forests are at the centre of a tangle of
conflicting interests. Accepting the fact that most of these interests are
legitimate is an essential first step if progress is to be made
in reducing today's rates of forest loss.

Facing the problem

It is simplistic and counterproductive to portray the complex issue of forest depletion as a struggle between the good forces of conservation and the evil of short-sighted greed and recklessness. People and governments in the developing world should not feel that they have to apologize for trying to find enough land to feed themselves and their families, for earning a living using the forests as humanity has always done or for promoting the growth of their national economies. The basic reason for the accelerating pace of forest depletion is the ever-increasing pressure on the world's finite resources.

Farmers must have land if they are to grow the food they need to survive. Many lack the financial resources or the knowledge required to farm their land on a sustainable basis; when its fertility declines they have no alternative but to move elsewhere. Others, displaced from their homes by famine, drought, war or inequitable distribution of land, move into new areas looking for land they can farm. Young people reaching adulthood on farms that are too small to feed them, let alone to enable them to marry and raise their own families, must go and find land elsewhere.

In much of the present developing world the forests and woodlands are the only source of new agricultural land. Many of those seeking land

Large dipterocarp logs in the Philippines are prepared for skidding by crawler tractor to a road, where they will be loaded onto trucks for transport to the processing plant.

they can farm are on the margins of society; they are usually desperate for the means to live. They are thus not easily deterred by laws, nor are they influenced by exhortations about the need to preserve the forests. Their clearing of the forests is not malicious or frivolous; they do it because it is their only way of surviving and the forest as it is has no value to them.

Loggers, whether they act as individuals, as small local companies or as large national or multinational corporations, see the extraction of forest timber as a legitimate way to earn a living. Harvesting forests for timber is as old as human society. The products are in demand in the local, national and international markets. In producing and selling goods and services that others require, those involved in the timber trade are no different from the rest of the human race. Condemned for being involved in cutting trees, loggers and timber traders ask why they are at the centre of public criticism, rather than the cultivators of rice, wheat, maize, beef and other foods who are mainly responsible for deforestation.

Many governments of developing countries deeply resent being cast in the role of wantonly destructive environmental villains. They have the unenviable task of trying to meet their people's needs in a largely indifferent world. They are faced with ever-worsening terms of trade as the prices of their export products decline and the costs of their imports increase. The burdens of debt that the developing countries have accumulated, often at the urging of the industrialized world, have suddenly been multiplied by factors beyond their control as world interest rates have soared during the past 15 years. Timber exports provide these countries with some of the foreign exchange they need to stave off today's crisis. Even when these governments want to manage their forests more effectively, most are short of the technical, financial and institutional capacity required to do so.

Nor are the conflicts over forests simply between those who would preserve the forests and those who would destroy them. Shifting cultivators see fertile soil where foresters see sustainable timber yields. Governments concerned with the broader interests of society find themselves trying to enforce regulations restraining the activities of farmers or timber companies so that watersheds or especially vulnerable areas are protected. Loggers find themselves confronting the people who traditionally depend upon the forest for their living.

Many of these competing claims

did not emerge in the past. The forests seemed an inexhaustible resource capable of satisfying the needs of all groups. Now it is clear that this was a faulty perception and the implications must be faced. Strictly preserving an area of forest for future generations means not only excluding squatters and encroachers, but also denying local people the right to earn a living from cutting and selling wood or hunting wildlife; it means opposing the attempts of a developing country's government to earn foreign currency. If, on the other hand, forests are to be conserved through sustainable management practices, then the people's needs for the goods and services of the forest can be met.

A slowing and eventual halting of the depletion of the world's forests is widely accepted as desirable; no one wants a world without forests. This can be done by harmonizing use and conservation through sustainable harvesting and cropping practices. But nothing will be achieved unless people's legitimate rights and interests are recognized. In some cases it may appear easy to take sides, but mostly it will be found that there are no simple or easy answers. Deciding on what should and can be done in each case will involve dialogue and almost always some degree of compromise and conciliation.

Taking an economic view

Most human activities involve costs and benefits that people have to weigh before they decide what they want to do. In general, no rational person embarks on a course of action if the costs are expected to outweigh the benefits.

In the same way, and at its simplest, the question to be asked in relation to the proposed management of an area of forest is whether the benefits of doing so exceed the costs. In purely financial terms, this involves a comparison between the income from the sale of timber and other products and the costs of management. The greater the revenues generated, the more likely it is that the management will be financially justifiable and will not require government or other subsidies. The more secure the financial base for a management system, the more easily it can be maintained on a sustainable basis, with the important proviso that from the revenue earned the required funds be channelled back into forest management.

The reality, however, is that most cases cannot be reduced to such simple terms. The gains and losses from conventional forest management are often distributed widely, with the gains going to one person, group or generation and the costs being paid by others. Moreover, the impacts may also be separated in time, with the gains typically coming immediately, leaving the costs to be borne in the future.

The further apart in space or time the costs and benefits are, the more difficult it is to reconcile them. Those clearing land for ranching are unlikely to be concerned with the contribution of their actions to future global warming or the consequences for another country that may be more prone to floods as a result of it. Nor will a farmer chopping down a tree for fuel on a watershed worry about adding to the siltation of a reservoir for hydroelectricity 100 km downstream. In this, the tree cutters are behaving no differently from the person in Los Angeles, Mexico City or Athens who takes a car to work and ignores the global warming implications of the journey or the contribution it makes to the smog from which the city is suffering.

Economics provides a way of objectively examining such issues and

of looking at the full range of costs and benefits involved wherever they fall. A logging contractor wondering whether to take up a lease, for example, will simply calculate if it will be profitable. An economic analysis will not only address the profit or loss of the contractor but should take into account the full range of benefits and costs to society.

On the benefit side, the economic analysis may, for example, look at the impact on local employment and any knock-on effects this has; the special benefits to the local or national economy from additional foreign exchange if the logs are exported; or the additional value produced if the forest land is being cleared for agriculture. On the cost side, the analysis may look at the effect on the future supply of forest products, the downstream impact if the logging is to take place on a watershed or any other negative impacts that the logging may have.

Similarly, if a decision is to be made on whether to permit the clearing of an area of mangroves for, say, a tourist development, an economic analysis could consider all the benefits and costs. The benefits would include the direct flow of earnings, the foreign exchange implications, the local employment generated and other beneficial impacts. The costs, in addition to those of clearing the mangroves and building and running the tourist complex, would include the loss of the timber and fish yield as well as any damage likely to be caused by the reduction in coastal protection or the loss of biodiversity.

In sum, forest management economics has to deal with much more than a short-term cash flow perspective, no matter where it is applied and no matter who is the owner or the manager. The economic assessment of management must integrate all the functions of the forest, and it is the inclusion of all the various functions that makes forest management economically sustainable. The main question then becomes how to make sure that the beneficiaries share the costs equitably.

This broader view of forest economics is particularly important in the tropics and subtropics, where the state is generally the owner of the natural forests. The financial costs and benefits are derived only from harvesting costs and prices for the limited number of tropical timbers that are marketed, which are strongly influenced by current market demand and rarely reflect the cost of their replacement. The state has to extend its evaluation beyond a limited financial analysis in order to meet the full range of its responsibilities to society. This wider analysis would encompass the broad spectrum of outputs of all goods and services from the forests and the effects external to the boundaries of the forests[1] and would compare all the costs and benefits of alternative forms of land use with natural forest management.

In considering the whole range of benefits arising from natural forests, whether inside the forests or external to them, it is difficult to determine monetary values for many of the costs and benefits involved. The tools presently available to assess and quantify the benefits and services provided by forests are inadequate. The valuation of services is particularly difficult. The forest manager is thus at a serious disadvantage when competing for scarce resources of funds or staff.

Comparing the management of forests with that of alternative forms of land use is equally difficult. The long time scales involved in forest management present a particular problem, because a great deal of conventional economic analysis is based on the reasonable observation that people consider immediate benefits to be more valuable than those that are deferred to some later date. A promise that $100 will be paid after ten years is obviously worth less than $100 immediately.

This is dealt with in economic calculations by discounting or reducing the value of costs or benefits that occur in the future. The discount rate is the percentage by which the value of future benefits or costs is reduced for every year until they are realized. A discount rate of 10 percent is commonly applied in the analysis of development projects. Using this rate, the value of a return or income of $100 to be paid in ten years is $39; it is just $0.84 in 50 years. If this analysis is applied to the relatively slow growth of timber under sustained yield management of natural forests, with a 50 or 100 year wait for a financial return, timber often turns out to be hopelessly unprofitable compared with other opportunities for financial investment, sometimes using the same land. In cases where other functions of the forest are not important and the alternative use of the land is equally sustainable, the most economically productive alternative may be the appropriate use.

Economists recognize, however, that evaluation of direct commercial costs and benefits is often not sufficient. The deeply felt human attachment to the care of the forest reflects appreciation of values other than the direct costs and prices. The traditional concerns for conserving forests are often associated with recognition of their protective functions in soil and water conservation and in provision of

shelter, which have material importance. There may also be cultural, aesthetic or spiritual values for which people care deeply. Economists seek to take account of these values. In the case of material benefits from protection, tangible values may be assessed from the cost of damage resulting from its absence. In the case of cultural values, the economic value is assessed as the value of the financial returns that are forgone when society decides simply to conserve the forest.

Another approach put forward is the idea that there is a stock of natural capital in the world on which the productivity and survival of the human economic system depend. This natural capital can be divided into two types: renewable, such as biomass, and non-renewable, such as petroleum and mineral resources. The moral assumption is that it is the responsibility of each generation to pass on an undiminished stock of natural capital to the one following. In an economy managed in accordance with this principle, it would be permissible to exploit renewable resources up to their sustainable limit. Non-renewable capital, on the other hand, could only be exploited at the rate at which substitutes were created. Although such theories are at an early stage in development and are highly controversial, they reflect a growing awareness of the urgency and complexity of the vast range of environmental and resource depletion problems facing the human race.

The Brundtland Commission expressed a similar idea in its description of sustainable development, which it said "is not a fixed state of harmony, but rather a process of change in which the exploitation of resources, the direction of investments, the orientation of technological development and institutional change are made consistent with future as well as present needs" (World Commission on Environment and Development, 1987).

The strength of economics, whether conventional or based on such ideas of sustainability, is that it provides a method for examining the range of costs and benefits associated with the choices open to society. Costs and benefits may be distributed in space or in time, and may be seen differently by those with different perspectives – the state, the private owner, the logger, local communities, environmental groups, etc. Benefits may be obtained as marketable goods or as public services, or even as other human values. Economic analysis permits an objective comparison of the overall costs and benefits of various options from various viewpoints. The data for reducing such complex calculations to purely cash terms are not yet fully available, but there are few cases where an attempt to examine the question from the economic viewpoint, assessing and comparing its overall costs and benefits, will not provide useful and illuminating insights.

If any type of forest is to be sustainably managed, the people and the countries concerned must be convinced that the land will remain more valuable under forest than under another form of land use. It will be necessary to find the means to assess the value of the forest accurately in monetary terms and to develop techniques to compare this value with that of alternative forms of land use. The real challenge of justifying the management of land as forest, however, will not so much be to make the return match the pre-established discount rate, but to ensure a fair distribution of costs and benefits that would make sustainable management a competitive proposal compared to other investment options.

[1]Examples of effects external to the forests, often called "externalities" by economists, include the contribution of forests to the stability of the global climate and the downstream consequences on farmland of clearing forests on a catchment area.

CHAPTER 5

Forest management options

Forest management can be defined as "deciding what one wishes to do with a forest, taking into account what one can do with it and deducing what one should do with it" (FAO, 1991c).[1] The objectives of management are represented by what one wishes to do, the physical and socio-economic context by what one can do and the prescriptions for the conservation and use of the forest by what one should do.

Forests can be managed in many ways and for many purposes. The technique used in each case will depend on the objectives, the type of forest, the available capacities and resources and the local conditions and constraints.

Clarifying objectives

The United Nations Conference on Environment and Development (UNCED) comprehensively described sustainable forest management. One of the key conference documents (UNCED, 1992) states that it encompasses the

> ...policies, methods, and mechanisms adopted to support and develop the multiple ecological, economic, social and cultural roles of trees, forests and forest lands...

the measures and approaches required at a national level to improve and harmonize policy formulation, planning and programming; legislative measures and instruments; development patterns; participation of the general public, especially women and indigenous people; involvement of youth; roles of the private sector, local organizations, non-governmental organizations and cooperatives; development of technical and multidisciplinary skills and quality of human resources; forestry extension and public education; research capability and support; administrative structures and mechanisms, including intersectoral coordination, decentralization and responsibility and incentive systems; and dissemination of information and public relations. This is especially important to ensure a rational and holistic approach to the sustainable and environmentally sound development of forests.

Such an encyclopaedic description risks itemizing every aspect of forest management but focusing on none of them. It is not a practical guide for individual cases. Forest management does not comprise everything desirable that might be done but is rather a matter of selecting and prioritizing the tasks that can and should be carried out for a particular area.[2]

Above all, it is important to be clear about exactly what the forest management is expected to achieve. If, for example, the required result is a large quantity of pulpwood, then the most obvious management option may be a plantation. If, on the other hand, the aim is to produce a wide range of forest products for local or international consumption, then natural forest management may be the most appropriate option.

Management objectives will vary substantially according to whether the forest is publicly, privately or communally owned. It has to be borne in mind that extreme objectives, although perfectly legitimate in their own right, are usually incompatible. Obtaining the highest possible sustainable timber yield will probably be incompatible with preserving the maximum biodiversity in the same location. It is only when there is a clear set of achievable objectives, analysed for compatibility and ranked in order of priority, that firm decisions can be made on the management methods to be adopted.

It is also necessary to accept the limitations to what can be achieved in any particular area. Because of the pressures of development and land scarcity, the forests of the developing world, in most areas, will not be left untouched. Management for sustainable timber yield, if it can be effectively carried out, can offer an economically productive alternative to clearing for agriculture.

Conservation of forest resources through sustained yield management is not the same as preservation through a policy of non-interference. In neither case can alteration be prevented. Ecosystems will continue to change even if left completely untouched by humankind. Management intervention in a forest, however, no matter how carefully or lightly carried out, inevitably alters the structure and ecology more quickly and in different directions than a policy of preservation; furthermore, if it is poorly carried out it runs the risk of causing serious and permanent damage. But sustainable yield management and total preservation are rarely the alternatives. More often the stark choice is between placing a forest under some form of management or clearing it for agriculture.

Management approaches

Forest management is not carried out in a vacuum. The method adopted must be appropriate to the physical conditions as well as to the socio-economic and institutional context in which it will be implemented.

Technical approaches that are suitable for the slow-growing temperate and boreal forests, with their limited number of species, may not be applicable to the much richer but often more ecologically fragile rain forests. The tropical dry forests, with their own special characteristics and vulnerability, will require yet another approach.

Forest management in Kenya; through the Green Belt Movement, millions of seedlings have been planted to combat soil degradation and an acute shortage of fuelwood.

The techniques adopted must be compatible with the technical and financial resources available. In the high-wage economies of the industrialized world, with their strong industrial infrastructures and relatively easily available investment capital, forest management tends to be heavily based on mechanization. In many parts of the developing world, where wage rates are low, labour abundant and investment capital extremely scarce, the management techniques of the industrialized world are likely to be inappropriate and even counter-productive.

Management must not be seen as something imposed from the outside by governments or forest services. Large areas of forest are under the management of the people living in and around them. Many such traditional management systems have worked well for long periods, and where this is the case the optimum approach may well be to avoid any external interference. When the traditional system appears to be breaking down or is subjected to new demands that it cannot meet, it does not necessarily mean that the system has failed completely and should be abandoned. The most effective approach will often be to reinforce the existing system or to help it to adapt to the new circumstances.

Where management has multiple objectives, depending on what they are, it may be possible to meet these simultaneously from particular areas of forest. The sustained yield of timber can be compatible with the social objectives of rural development. Protecting a watershed is usually compatible with the conservation of wildlife. In other cases, where such multipurpose management is not possible, the best alternative may be to set aside areas of forest to be managed for different purposes. One part may therefore be managed intensively for timber, another part may be left under the control of local people for the less intensive production of other outputs such as fodder or medicines, while other areas may be conserved for ecological or other environmental reasons.

There is no universal management prescription. What is essential in all cases is clarity about objectives and about who is responsible for pursuing them and under what conditions. Only when these are properly defined and established and priorities are assigned is it possible to develop a strategy that will enable the objectives to be realized.

Forest management in progress

Once the management objectives have been clarified, the task is to decide on the management and silvicultural techniques to be used.

Temperate and boreal forests

The main management focus in the temperate and boreal forests is upon sustained timber yield. In many areas, this is carried out in a manner that is compatible with the use of forests for watershed protection, recreation and other uses. In recent years, increasing attention, especially in the United States, has been focused on issues of biodiversity and the conservation of endangered species and ecosystems.

Methods of managing the temperate and boreal forests for sustained timber production have long existed. Increasing the area of these forests under such management depends more on political commitment and available funds than on acquiring new knowledge about what needs to be done.

Many of these forests are comparatively simple in structure and species composition. This is particularly the case in Europe, where selection and management over centuries have increased the concentration of timber-and wood-producing species and the degree of uniformity in the forests.

Because of the relatively high density of commercial trees, especially in forests that have been under long-term management, most of the trees are cut during harvesting. Conifer forests are generally clear felled. In broadleaf or mixed forests, however, there is an increasing tendency to carry out selective felling.

There are well-tested methods for ensuring that regeneration takes place after harvesting. These are recognized as good forestry practice and are generally accepted and implemented by the timber industry. They include taking steps to guarantee natural regeneration and sowing or planting of seedlings. In other cases, an area may be completely cleared and replanted.

The temperate and boreal forests are not only managed for timber; cork, naval stores[3] and mushrooms are examples of important commercial products from temperate forests under sustainable management, and many of the temperate forests in industrialized countries are managed for recreation or amenities as well as for forest products.

In general, the temperate and boreal forests are fairly robust systems as far as forest management practices

are concerned. The soils are not particularly fragile and are not endangered by logging, provided that the proper precautions are taken over the construction of roads and that felling is avoided on the steeper slopes. The threats to the temperate, and possibly to the boreal, forests arise more from the effects of pollution and fire than from mismanagement.

Moist tropical forests

The main focus of concern over the moist tropical forests is their rapid rate of depletion. Most of this is caused by clearing for agriculture and is therefore outside the scope of forestry management or silvicultural remedy. Questions of rural development and land-use planning are involved as well as decisions and action by a broad spectrum of authorities and expertise both inside and outside the forestry sector.

Practical management interventions have been limited in most countries, and only a small proportion of the tropical moist forests is subject to effective management. In most other cases the only activities are supervision of logging in support of royalty collection.

Sustainable yield management offers the prospect of realizing the economic potential of the moist tropical forests while at the same time maintaining their basic structure as well as their capacity to deliver a variety of other goods and services. It is, however, considerably more difficult than in the temperate and boreal forests.

The tropical moist forests are inherently complex systems. Because relatively long rotation periods are involved and because sustained yield management is comparatively recent in most areas, some uncertainties about optimum techniques and exact yields and outcomes inevitably remain. The indications are, nevertheless, that management of the great majority of these forests for sustained output of wood is technically feasible.

The main obstacles to the more widespread application of sustainable yield management are economic. Far from being a uniform and bountiful source of easy financial gain, as is often popularly supposed, the tropical moist forests tend to share a number of features that make profitable logging, let alone sustained yield management, difficult to achieve.

The problems of managing these forests include the large number of species, many of which are non-commercial; the rapid and luxuriant growth of vines and creepers in the open spaces created by felling; the general fragility of the soils and their vulnerability if fully exposed; the difficulties of access; difficulties in inducing natural regeneration; and the extremely arduous working conditions, especially in the wetter tropics. Harvesting operations are therefore expensive, and logging companies tend to demonstrate strong resistance to any increase in logging expenditures to reduce damage to the remaining stand or to the soil, especially when the companies involved are small and undercapitalized and the prospects for long-term investment are unfavourable.

The first step towards the sustainable management of an area of forest is to draw up a management plan. A forest management plan prescribes the activities required to meet its objectives. The prescriptions may be simple or complex. A detailed plan requires a considerable

amount of information such as an inventory of the standing stock and of its condition and age or size composition, as well as an assessment of the soils, slopes and other factors that affect the way in which the silvicultural and logging operations should be carried out. This demands a substantial amount of work which may simply be impossible for a poorly funded and understaffed forest service. But even a simple working plan, based on cautious assumptions, can provide a basis for sustainable yield management.

The working plan lays down the conditions that logging concessionaires are expected to follow. Access roads are supposed to be constructed in a way that minimizes forest damage and the risks of erosion. The numbers, species and sizes of trees as well as the methods of felling and extraction are specified to limit the damage to the remaining forest. It is particularly important to prevent "creaming" of the forest, in which only one or very few species with the best form are removed, making it more difficult to manage the area on a sustainable yield basis.[4]

The design and construction of roads is critical. They may occupy up to one-sixth of the total forest area in European forests, although less in tropical forests. Roads can be a major cause of forest damage. One estimate is that 90 percent of the soil erosion associated with logging is directly attributable to roads. Studies have shown that when proper attention is paid to the overall planning of the road layout and logging operations, fewer saleable trees are left behind, there is less damage to other trees during felling and there are fewer accidents. Costs are lower and the amount of damage to the forest environment is substantially reduced.

Harvesting should be seen as a silvicultural operation, linked to the initial inventory. The most critical element in the implementation of the working plan is therefore the degree of control and supervision exercised over logging operations. This is a major area of weakness in most developing countries. When foresters are poorly trained, ill-equipped and underpaid, the control exercised over logging tends to be weak at best and is often absent. It is hard to expect otherwise when foresters may have to depend entirely on the logging companies for their transport, accommodation and living requirements. Where there is no commitment to the enforcement of regulations in political circles and the upper ranks of the forest service, logging is effectively uncontrolled and drawing up working plans is more or less meaningless.

The final management task is to ensure that the forest regenerates in a manner that will permit it to be reharvested in due course. Monocyclic management methods are based on a single harvesting after which the forest is left until ready again. In polycyclic management systems, there are intermediate harvests in which the faster-growing trees are extracted. Within these broad categories a wide variety of techniques are used, based upon the information provided by the first and subsequent inventories.

One of the main difficulties in putting into effect management for sustained yield in tropical forests is the small proportion of commercially valuable species that are harvested. This is at first sight paradoxical, but is explained by the fact that once the relatively few commercial trees have been removed, the open areas are quickly covered with creepers and are invaded by other pioneer species. If the desired species are not pioneers then their natural regeneration in the face of this intense biological competition is slow and uncertain. The management challenge is to find ways of manipulating the natural processes of forest regrowth so that the commercially valuable species are encouraged to regenerate and preferably so that their numbers relative to the non-commercial species are increased.

Sometimes the area being managed is enriched by planting seeds or saplings of the commercial species in the logged area. This requires a considerable amount of labour in planting and weeding and is expensive. A cheaper alternative is to rely on natural seeding, when possible, and on improving the growth of the younger commercial trees by techniques such as liberation thinning – the elimination of competing species in the immediate vicinity of immature commercial trees, leaving areas where there are no commercial trees undisturbed – and climber cutting.

Whether natural regeneration or enrichment planting is used, measures are usually taken to reduce the competition to the desirable seedlings and to the advance growth. Formerly these objectives were met by killing some of the surrounding non-commercial trees either by girdling or by poisoning with arboricides such as sodium arsenite. The elimination of trees is becoming rarer as experience shows that species regarded as non-commercial at a particular time may become saleable within a decade or two. Liberation thinning reduces disruption of the forest structure as well as costs.

A particularly effective technique for enrichment planting was developed in Uganda during the 1960s. Charcoal makers were allowed into the logged forest areas and were encouraged to convert those species not used for timber into charcoal. This opened up the forest for the growth of the commercial species, which were planted in the cleared areas. Using the trees unwanted by the timber trade for charcoal avoided the cost and waste involved in poisoning while providing employment and meeting fuel needs.

A method developed in Côte d'Ivoire during the 1930s involves limited clearing of the forest after logging, followed by close planting of desirable species. As these grow, the non-commercial trees are gradually poisoned around them. This gives good results but requires a considerable amount of labour. Another management system, begun in Trinidad in 1926, was originally based almost entirely upon regeneration by planting but has been gradually changed in the light of experience and now relies completely on natural regeneration.

An experimental technique at present being developed and tested in Peru is particularly interesting in that it involves the active engagement of the local community. Although clear felling is used, it is only carried out in narrow strips running along the contours of the land. This reduces the danger of damage from erosion and facilitates regeneration from the adjacent uncut forest. To maximize local employment, the logging operations are mainly based on hand labour and the use of oxen, although in some cases portable sawmills are used on site to improve the rate of utilization and the marketability of the timber.

The management system chosen and the degree of intensity with which it is applied will vary depending on local conditions and the objectives. If maximum production of commercial timber is the aim, one of the more intensive management systems or complete replanting after logging is likely to be the most appropriate. Where a limited harvest of top-quality timber for veneers and other high-value uses is the objective, less intensive methods may be best.

In all cases, a critical requirement is that the management method chosen should be related to the implementation capabilities available. This requires a realistic appraisal of the technical and managerial skills available, the financial resources obtainable and the degree of political and administrative commitment to forest management. There is no point in adopting a highly elaborate management plan if there is no way it can be carried out. It is far better to have a simple system that provides

some degree of control than an ambitious one that exists only on paper.

The management approach instituted in the Congo, for example, has been deliberately chosen to match the size of the existing staff of the forest service. Under the system, the forest service has simply defined the areas that can be logged and has set conservative limits on the numbers, sizes and species of trees that can be taken from them. Control is exercised by inspection of the logs being extracted. No silvicultural treatment is involved, and road construction and harvesting practices are subject to spot checks. Once logging has taken place, access is restricted until the area has recovered. The system is geared to the capacities available and, in practice, provides a substantial amount of control. As the human and material capacities of the forest service increase, the control system can be improved and silvicultural treatments can be introduced.

This section has concentrated on the management of moist tropical forests for the production of timber. There are, however, many other products harvested from moist tropical forests. Rattans are of great economic importance for Indonesia (which produces 80 percent of the world's annual consumption), Malaysia, the Philippines and several other countries. Cardamom (*Elettaria* sp.) is grown in southern India under the shade of tropical moist forest. Another important non-wood forest product from India (albeit from the moist deciduous forests) is lac, which comes from the protective covering of an insect that infests the young shoots of *Schleichera* spp., *Zizyphus* spp. and other host plants. From the forests of Brazil come natural rubber, gums, nuts (including the Brazil nut, *Bertholletia excelsa*) and palm hearts.

Details of management techniques for mangrove forests have not been considered, but the basic principles of forest management apply also to these types, which provide, in addition to wood products (such as timber, fuelwood, poles and pulp), also a great range of non-wood forest products which include tannin, nipa palms, shellfish, honey, crocodiles (from crocodile farming), agriculture, salt, ecotourism and coastal protection (FAO, 1993i).

Dry tropical forests

The dry tropical forests pose management challenges very different from those of the moist tropics. Most of the native tree species are slow growing and drought tolerant. During hot dry spells, biological activity is reduced to a minimum as a means of survival. Fire is an important hazard.

The wood produced is usually hard and durable and with few exceptions is generally only commercial on a local basis. Regeneration often relies on grazing animals eating the pods

An acacia woodland in Kenya; many important non-wood products come from the continent's dry forests.

and excreting the seeds; otherwise the seeds are able to survive for years in the soil. A high proportion of the tree species coppice, producing vigorous new growth when the main trunk is cut. Many of the species are fire resistant when larger than pole size.

Wildlife is a significant element in the management of these areas. It is extremely important at a local level as a source of meat and other animal products. It is also of major significance in the tourist trade of countries such as Kenya, the United Republic of Tanzania and Zimbabwe.

Where rainfall is scarce but reliable, sustained yield management is technically feasible. This is usually based on replacement or enrichment planting. The drier the area or more erratic the rainfall, the poorer the record of replacement planting tends to be. In some areas, studies have shown that the yield from exotics may be less than that of the indigenous forest cleared to make place for them.

The management emphasis in the drier areas has consequently been shifting towards the regeneration management of existing forests with indigenous species and the forestation of degraded or even completely barren areas. In a number of countries, demonstration plots in which cutting and grazing have been forbidden and fire has been excluded for a number of years have shown a remarkable ability to regenerate both from coppice and from seed that has lain dormant in the soil. There is no doubt that this suggests a management method capable of restoring and sustaining the productive capacities of large areas of forest in these areas.

Important non-wood products from dry forests include gums (such as gum arabic), fodder, honey and grazing, whose production is included in the objects of management of some forests. The management objectives of other forests include maintenance of populations of wild animals, either for hunting or in support of the tourist trade.

The main problem in implementing forest management schemes in most of the dry forest areas is the intensity of existing land use. Even in badly degraded areas, people may rely completely on what is left of the forests for browse and fuel. Closing off areas for regeneration, even though it will produce long-term gains, can impose intolerable short-term burdens upon people. Where lands are in common ownership there may also be difficulties in arriving at satisfactory methods of sharing out the various benefits and costs involved.

*Conserving genetic resources
and biodiversity*

This section concentrates on the conservation of plant biodiversity and plant genetic resources, but in fact the same principles apply to the animals in the forest ecosystems.

Although conserving biodiversity and conserving genetic resources are frequently seen as the same thing, they are different in that action is focused at different levels: within the ecosystem or within the species. It is, for example, entirely possible to conserve an ecosystem or a species in it but to lose genetic diversity; likewise, it is possible to conserve genetic diversity while the number of species in the ecosystem may decrease.

Conservation of genetic resources aims at ensuring that the widest possible range of genetic variation for a given target species is identified and

conserved. This will usually include surveys, demarcation, management and monitoring of conservation areas and the collection and storage of seeds or tissue. Management may imply the elimination of competitors or the opening of the forest canopy to promote the growth of the target species.

On-site conservation of genetic resources requires planned and systematic management of clearly defined target species in a network of conservation areas and managed resource areas. It should be aimed at the maintenance or enhancement of the within-species variation found in these species. The primary challenge for the conservation of genetic resources is thus not to select, set aside and guard protected areas containing genetic resources; rather it is to maintain the genetic variability of the target species within a mosaic of economically and socially acceptable land-use options. These may include strictly protected areas as well as multiple-use reserves, managed forests and agro-ecosystems. While, in general, the conservation of genetic resources outside protected areas in places used for wood production (for example) will require more intensive management and the monitoring of target species, such areas are not inherently less compatible with genetic conservation of the species being harvested in a sustainable manner, or of associated species, than are strictly protected areas. It should thus be possible to harmonize conservation of genetic resources with sustainable use of much of the land area of a country by including concerns for genetic conservation of target species as a major component in land-use planning and management strategies (FAO, 1991a).

Conservation of biodiversity is concerned with the full range of plant and animal species and their interrelationships. The objectives can easily become blurred, since natural ecosystems are in a constant state of change in which new species are evolving or being introduced while others decline. For instance, the number of species in the climax forest is considerably less than in intermediate successional stages. It is therefore important to specify whether the management system is intended to attempt to freeze biodiversity at its present level, to achieve a different balance or to allow nature to take its course.

Most national parks, to take a practical example, are protected against forest fires. This reduces the competitive advantage of pioneer and fire-resistant tree species. Over time, the balance of tree species, as well as the populations of birds, mammals and other animals, changes. The forest will tend to become more vulnerable to fire damage as the undergrowth thickens and the number of old and decaying trees increases. In this case, protection against forest fires, which may appear to preserve biodiversity, is an interference which will ultimately result in an "unnatural" forest.

Thus, clear objectives are necessary in developing strategies for conserving biodiversity, just as they are in the management of forest areas for any other purpose. Simply protecting an area may result in a reduction in biodiversity and the elimination of plant or animal species that are unable to adapt to the inevitably changing conditions of today's world. Protecting endangered species or maintaining what is felt to be a desirable balance between species may require substantial outside interference in the natural working of the ecosystem.

In all cases, the assent and cooperation of local people will be required if the management of forest areas to conserve biodiversity and genetic resources, however these are defined, is to be successful in the long term. Attempting to drive people out of areas on which they rely for food, fuel or other goods is almost certain to fail. Local people must be involved in the initial planning and final implementation of the management system and must derive tangible benefits from it. Where people are required to give up existing benefits or access rights, they must be adequately compensated.

An example of the conservation of biodiversity is the establishment of a 77 000 ha national nature reserve in Niger. The reserve was proposed in 1982 and was formally established in 1988. The local inhabitants were consulted at all stages to ensure their cooperation. A wildlife sanctuary occupying 12 percent of the area was created in a part of the reserve rarely visited by people. The establishment of the reserve was associated with a rural development programme which provided health training, adult literacy classes, help with well digging and other assistance to the community. Another example is from Peru, where a village community from an area west of the Huascaran National Park has established an *in situ* conservation area of 0.15 ha of *Alnus jorullensis*. A nursery has been established which produces this species for planting in association with agricultural crops to fix nitrogen and to provide fuelwood and soil protection.[5]

Trees in the landscape, agroforestry and urban forestry

Sustainable forestry management is not only concerned with trees growing in large blocks but in its broadest sense involves trees growing in the rural landscape, including those on agricultural land, and trees in urban areas.

In most settled farming societies, trees in the landscape provide a number of products (for example, fuelwood, poles, animal fodder, fruit and fibres) and perform several important functions (such as shading, protecting from the wind and decreasing soil erosion).

The numbers of trees in the rural landscape are, however, falling in many areas. Among the immediate causes are shorter fallow periods, agricultural mechanization, increasing grazing pressures which prevent natural regeneration, the breakdown of traditional communal farming systems and increased fuelwood cutting – especially that carried out to meet urban demand. These increased pressures on wood resources reflect the complex and rapid social, economic and cultural changes taking place in the societies involved.

In attempting to promote increased tree growing by farmers or by rural communities, it is therefore essential to understand why trees are disappearing, why they are not being replaced and what the local needs are that trees can fulfil. Many of the early tree growing campaigns failed because they focused almost exclusively on growing trees for fuel, which is rarely a priority among rural people. People are often more likely to plant or manage trees for the production of higher-value products, such as timber, poles, fruit or fodder, or even for a service, such as boundary demarcation, shade or ornament, with fuel regarded as a by-product.

Various approaches are now being used in different parts of the developing world to encourage tree growing by farmers. They include educational campaigns, the provision of free or subsidized seedlings, training for farmers in setting up commercial nurseries and support for NGOs that promote tree growing. The results are mixed, indicating that in many cases the campaigns are still not addressing real concerns or that problems of land tenure and tree ownership, which are often the underlying cause of the depletion, have not been resolved.

Agroforestry represents the integration of agriculture and forestry to increase the productivity or sustainability of the farming system and/or to increase farm income. The definition adopted by the International Council for Research in Agroforestry (ICRAF) is as follows: "Agroforestry is a collective name for land-use systems and practices where

Agroforestry can increase farm productivity and income; this experimental plantation in Tanzania consists of maize successfully intercropped with Pinus patula.

woody perennials are deliberately integrated with crops and/or animals on the same land management unit. The integration can be either in spatial mixture or temporal sequence. There are normally both ecological and economic interactions between the woody and non-woody components in agroforestry."

Agroforestry includes a wide variety of land-use systems, ranging from those in which trees are planted and managed on agricultural lands to those in which agriculture is practised on forest lands without resulting in deforestation. Many traditional systems have been developed over time through trial and error by farmers and reflect local environmental and socio-economic conditions. The home gardens of Indonesia are one example, while another is the maintenance of naturally occurring *Faidherbia albida* (formerly *Acacia albida*) in agricultural fields in semi-arid Africa. The contribution of these trees to agricultural production is widely recognized, and these trees are often protected by traditional laws.

Considerable work has been done over the last 15 years to develop agroforestry systems, by devising approaches for research and extension, and to test new systems. Agroforestry, however, is still a new science, and far more has yet to be done to develop systems that are not only biologically and economically viable but socially acceptable as well.

The integration of trees into agricultural systems is not the only way in which trees are incorporated in the landscape. Trees have long been used to beautify and ameliorate the urban environment, but in view of the prediction that half of the world's population is expected to be living in urban areas by the year 2000, the need for urban forestry[6] is increasingly recognized. For instance, the green cover (trees and shrubs) of the city of Beijing has increased from 3 percent in 1949 to 26 percent today, or about 6 m^2 per person (Dembner, 1993); in contrast, the green cover of Mexico City is 2 percent, not quite 2 m^2 per person (Cabellero Deloya, 1993). Both cities are aware of the benefits of urban forestry and are attempting, despite considerable difficulties, to increase the area of green cover.

The role of trees in cities lies in the provision of aesthetic and environmental benefits [the latter including climate modification, energy conservation, noise level reduction and improvement of air quality (see e.g. Nowak and McPherson, 1993)] and possibly in the provision of wood products and foodstuffs. The development of techniques for the sustainable management of urban trees and forests will, like the sustainable management of forest plantations, be complementary to the development of sustainable management strategies for the world's natural and semi-natural forests.

Protecting forests against fire, insects and disease

Fire, insects and disease are integral to forest ecosystems. However, under certain conditions they can cause widespread damage and disrupt the flow of goods and services that forests provide. They can affect the growth and survival of trees, water quality

Insects and disease destroy millions of cubic metres of forest products such as sawtimber and pulpwood.

and yield, wildlife habitats, species diversity, the supply of forage and fodder and recreational, scenic and cultural benefits. Consequently measures to protect forests from fire, insects and disease are an essential part of forest management if sustainable levels of goods and services are to be assured.

Natural fire has influenced plant communities over evolutionary periods of time. In semi-arid regions, where fires are frequent, forests and woodlands have evolved with fire and plants have developed adaptive traits which ensure their survival or enable them to compete with less fire-tolerant species.

Fire can also be an agent of destruction; a low ground fire burning in a forest of fire-tolerant trees will kill regeneration and understorey vegetation. Fire-tolerant trees can sustain injury from fire which makes them more susceptible to attack by insects or fungi. More intense fires can kill all the vegetation on a site. Years of growth may be destroyed in a matter of hours and endangered habitats may be lost. Destruction of vegetation by fire may cause soil erosion, especially on steep slopes, leading to siltation of water supplies. Human property and life may be lost. In addition, when forests burn, carbon that is stored in woody tissue is released into the atmosphere as carbon dioxide and other greenhouse gases. Increases in atmospheric levels of these gases is causing concern because of their potential influence on global climate.

Fire is also a valuable tool in land management. It is used to prepare land for agriculture. Much of the tropical deforestation and associated burning are done to support shifting cultivation, which is practised by some 200 million people on 300 to 500 million hectares worldwide. Fire is an important silvicultural tool in forestry operations and is used for slash disposal, fuel reduction and the preparation of sites for planting or natural regeneration.

It is estimated that 12 to 13 million hectares of forest and other wooded lands are damaged annually by wildfire. When conditions are favourable, catastrophic wild fires can occur. In 1982–83, following a severe drought, some 3.6 million hectares of primary and secondary rain forest were destroyed in East Kalimantan, the Indonesian portion of the island of Borneo. In 1983, the "Ash Wednesday" fire burned over 340 000 ha in southern Australia; it destroyed 300 000 farm animals, damaged 2 500 homes, injured 3 500 people and killed 75. During 1987, over one million hectares burned in northern China, and in 1988, fires in Yellowstone National Park in

Fire may be used as an effective forest management tool but, if not controlled, it can be highly destructive.

Montana and Wyoming in the United States burned over 320 000 ha.

Wildland fire management is an essential element of sustainable forest management and consists of prevention, presuppression planning and suppression. Measures to prevent forest fires focus on education, legislation, the reduction of the volume of combustible fuels and the construction of fire breaks. For suppression, resources have to be provided for fire-fighting equipment and the recruitment and training of fire-fighters.

The role of natural fire in the dynamics of forest ecosystems must be considered when designing fire management programmes. Exclusion of fire from ecosystems where fire has a natural role can result in changes in vegetation, accumulations of fuels, increased risk of pest

The larval stage of the pine caterpillar, Dendrolimus punctatus, *a major defoliator of pines in China and Viet Nam.*

outbreaks and, ultimately, the risk of a catastrophic fire.

Insects and disease can also cause extensive forest damage. Every part of a tree – the foliage, buds, flowers, seeds, stem, bark, wood and roots – can serve as host material for these agents. Trees of all ages, from seedlings to mature trees, are subject to attack. In addition, insects and pests affect logs and wood products.

The ecological, social and economic impacts of insects and disease are far-reaching. Pest activity can significantly reduce yields of wood products. In the United States, bark beetles alone kill trees with a volume of 25.5 million cubic metres of sawtimber and pulpwood annually. The yield of fruit, mast or other food products on which wildlife, livestock or humans may depend can also be significantly reduced. Defoliating insects cause a decrease in the availability of fodder derived from hardwoods in India and many other countries. Recreational and scenic values of forests are reduced where extensive areas of forest have been damaged by defoliating or tree-killing insects, and trees weakened or killed by insects and disease in forest recreation areas are a safety hazard. Dead trees left behind in the wake of pest outbreaks increase the volume of combustible fuels. Consequently, when a fire occurs in pest-damaged forest it will burn with greater intensity and will be more destructive and more difficult to extinguish.

In the United States, fungi that cause decay in living trees account for more loss in sawtimber than fire, insects, weather or any other disease agent. Decay of wood products is also significant. Approximately 10 percent of the timber harvested annually in the United States is utilized to replace wood that has deteriorated because of decay caused by fungi. In the tropics and subtropics, insects such as termites are capable of causing extensive damage to living trees, homes and other wooden structures.

Pests that are accidentally introduced can eliminate a host plant from an ecosystem and reduce species diversity. Chestnut blight (*Endothia parasitica*) eliminated American chestnut (*Castanea dentata*) as a major component of the hardwood forests of the eastern United States. This tree was once highly prized for its attractive, decay-resistant wood and for its nuts which were used by wildlife. More recent accidental introductions of insects such as cypress aphid (*Cinara cupressi*) in eastern Africa, Leucaena psyllid (*Heteropsylla cubana*) in Asia and the Pacific and European pine shoot moth (*Rhyacionia buoliana*) and a European wood wasp (*Sirex noctilio*) into South America threaten the future viability of large areas of exotic, fast-growing plantations.[7]

The natural defence mechanisms of trees and the natural enemies of pest species usually keep the damage to trees at low levels. When trees are weakened by drought, inadequate management, pollution or other factors, the risk of pest damage increases. The decline currently affecting many tree species in Europe and North America is believed to be the result of a combination of these factors. Single-species plantations and natural forests in the boreal and temperate zones which contain only a few species tend to be more susceptible to build-up of pests than tropical forests, in which several hundred plant species may grow on a single hectare.

The protection of forests from attack by insects and disease is best addressed under the concept of

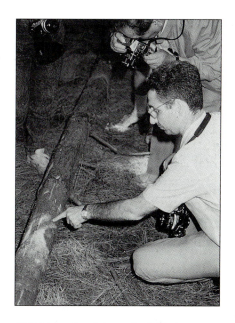

Field trips and workshops attended by professional foresters help to create an awareness of pest problems and facilitate the dissemination of information on pest management.

integrated pest management (IPM). IPM is a process of decision-making and action that takes into consideration the ecology of pests and their hosts and the ecological, social and economic consequences of pest damage and pest management actions. Emphasis is placed on pest monitoring, on understanding the underlying causes of outbreaks and on maintaining or improving the health of forests rather than on controlling pests.

IPM involves the use of one or

more tactics to reduce losses caused by pests to a level that permits immediate and future management objectives to be met. These tactics include silvicultural action to prevent pest build-up, regulation to prevent pest introductions, the use of predators, parasites and diseases to reduce pest populations, and genetic selection of pest-resistant strains.

Concerns about the potentially negative side effects of chemical pesticides, such as environmental contamination and hazards to human health, have made their use less desirable. In addition, chemicals are expensive and are often not economically justified, especially in developing countries. Despite these drawbacks, chemicals are still considered a part of IPM, but only if they are used selectively and as a last resort, and short-lived compounds are preferred.

IPM is a dynamic process under which new approaches and new technologies are continuously tested and evaluated. Those that prove effective are quickly integrated into ongoing programmes.

Plantations

Plantation forestry is a well established form of intensive forest management. It is estimated that plantations provide 7 to 10 percent of the world's present commercial wood

Plantations of the fast-growing conifer Cupressus lusitanica *can produce up to 300 m³ per hectare of timber in less than 30 years.*

production. There are roughly 100 million hectares of forest plantations in the world (Gauthier, 1991).

Statistics on plantations must be treated with caution. Sometimes the figures are based on the accumulated area planted without any deductions for the areas felled. When estimates are made of the areas planted by local communities or farmers, the margin of error is even larger. Often the figures are based simply on the numbers of seedlings distributed and not on the numbers planted or surviving. On the other hand, the figures may omit the numbers of trees planted by farmers from their own seedlings.

The net area of forest plantations (taking into account the estimated survival rates) in the tropical countries in 1990 was estimated to be about 30 million hectares, counting industrial and community plantations but not including trees planted by farmers themselves on their own lands (Pandey, 1992). It has been calculated that plantations were established at an average rate of around 2.6 million hectares per year during the 1980s, and the present rate may be as high as 3 million hectares per year.

The world's rubber, coconut and oil palm plantations are not included in the area of forest plantations. These are mainly in Asia and the wood obtained from them is increasingly important. Their total planted area is about 14 million hectares, of which about 7.2 million are planted to rubber, 4.2 million to coconut and 2.7 million to oil-palm.

One of the oldest methods of plantation management is the taungya system. This was introduced in Myanmar (then Burma) in the 1850s for teak plantations and has since been used in many other areas. After logging, the forest is replanted. Farmers are allowed to grow crops in the planted areas in exchange for protecting the saplings and weeding the plot. In a few years, when the

Forest plantations supply up to 10 percent of global commercial wood production; however, not all plantations are successful, and care must be taken both in the planning and management of new projects.

new trees are so large that food crops can no longer be grown in the shade they cast, the process is repeated in another area. Although the system offers a low-cost means of plantation establishment, it is not necessarily sustainable, since it relies on a shortage of permanent agricultural land and a supply of people willing to take part.

Plantations can be highly productive. The increment of timber from a tropical plantation may be 30 m^3 per hectare per year (or more) compared with 2 to 8 m^3 from a managed natural forest.[8] But such figures need to be treated with caution. Experience shows that the yields assumed at the planning stage of many plantations are overestimated, often by a factor of two or more. This was the case, for example, in the dendrothermal programme in the Philippines, which intended to use wood from rapidly growing plantations as fuel for electric power stations. The fact that yields turned out to be half those planned or less was one of the reasons for the failure of the programme.

Other aspects of plantation forestry, especially in the tropics, also leave much room for improvement. There are numerous examples of plantations that have failed or sites that have been degraded by ill-chosen exotic species. A review of tropical

plantations (Pandey, 1992) observed that planning is generally poor, particularly in relation to vital issues such as the matching of species to the site. It also noted that plantation projects are often designed in haste, with scant attention paid to important issues because of time or financial constraints.

In the developing world, the main physical constraint to the future contribution of plantation forestry is the availability of land. With expanding farming populations using all the available unforested land for food production, the areas available for plantations are becoming ever more restricted. The experience of the past two decades has shown that degraded or "waste" land may be the only resource available to poor people.

There are, however, large areas where the natural forest has been badly degraded or where the soil fertility has been lost through overcultivation, which could be used for plantations. Such schemes could provide a source of employment and long-term income, provided the existence and needs of local people are recognized from the beginning. There are also large areas of salt-affected land; the total in the developing world is estimated to be about 150 million hectares. Much of this land could be brought into productive use by planting it with salt-tolerant trees. But even here there could be competition from agriculture; major efforts are being made in some areas to rehabilitate these lands or to use them for salt-tolerant crops.

As investments, the main feature of plantations is their long production cycle. Tropical timber plantations have rotations ranging from 15 to 20 years up to 60 to 80 years. In the temperate zones, the rotation periods are even longer. Because broadleaf species such as oak are now being used in response to the popular reaction against uniform conifer plantations in some European countries, some of the plantations now being established will have a rotation period of up to 150 years.

A highly ambitious global forestry programme was included in the Noordwijk Declaration (Noordwijk Ministerial Conference, 1989), the resolution of a ministerial-level, global meeting convened to discuss the problems of atmospheric pollution and climate change. Among the aims set out in the declaration were a world net forest growth of 12 million hectares per year by the turn of the century. It was not anticipated that all of the net forest growth would come from forestation of currently treeless land, but nevertheless a high proportion of the increase would be new plantations.

Eucalypts have been widely planted in tropical and subtropical regions; this plantation of Eucalyptus globulus *in Chile is being used for production of mine timbers.*

At present, there appears to be little prospect that the required planting rate will be achieved

Plantation forestry nevertheless can continue to make a significant and expanding contribution to the needs of the world for wood. Without plantations, supplies can only be drawn from the natural forests. It is therefore important that governments continue to invest in plantations. It is also highly desirable that private-sector investments be encouraged. Because the returns from plantations are subject to a variety of long-term risks ranging from price collapses to natural calamities, as well as having long payback periods, it is essential that governments provide the necessary incentives to communal and private investors – with proper controls.

Plantations generally have one major objective for their establishment; often it is the production of wood, but some plantations are created for shelter and protection. Sometimes, however, plantations can supply both wood and non-wood products. Examples are the growing of cardamom under the shade of eucalypts in India, or the raising of mushrooms in *Pinus radiata* plantations in Chile (FAO, 1993a).

There can be no pretence that plantations can provide the full range of goods and services supplied by a natural forest. They comprise tree crops, analogous to agricultural crops, with a simplified ecology of one or at most a few species usually chosen for their yield and ease of management. The primary purpose of most plantations is to produce wood or other products quickly and cheaply. Their role, which is a highly valuable one, is as a complementary element in national or global forestry management strategies.

[1] The following is quoted from the same paper (FAO, 1991c): "In the broadest sense, forest management deals with the overall administrative, economic, legal, social, technical and scientific aspects involved with the handling of conservation and use of forests. It implies various degrees of deliberate human interventions, ranging from action aimed at safeguarding and maintaining the forest ecosystem and its functions, to favouring given socially or economically valuable species or groups of species for the improved production of goods and environmental services."

[2] "Forest ecosystems are managed for a variety of objectives related to the many goods (wood and non-wood products) and services (e.g. soil and water conservation, conservation of biological diversity) which they can provide. In most cases they are managed for more than one objective. Even in the case when a single objective is declared, forest management may achieve other purposes which may or may not be clearly expressed. For instance, if sustained timber production is the only explicit objective, forests continue to provide other services such as those of soil and water conservation.

"Prioritization of the objectives of multipurpose management is needed in order to facilitate the choice among the conflicting demands put on the forests under management. In particular, one main objective must be given priority over the others. However, in striving to achieve this objective, forest managers must see to it that all the other objectives remain fulfilled, at least partially" (FAO, 1991c).

[3] The term "naval stores" refers to the products of the resin industry in the United States, particularly to turpentine and rosin but also to pine tars and pitch.

[4] "While harvesting of mature trees of good quality is among the stated objectives of forest management aimed at the production of timber, pressing market demands coupled with inadequate forest management practices may lead

to highly selective harvesting having negative (dysgenic) effects on the future development of the stand. Silviculture rightly calls for harvesting of 'the best', but this must not be done without due consideration to regeneration potential and the quality of the next generation crop" (FAO, 1993b).

[5]"West of the Huascaran National Park the community Ramon Castilla has established an *in situ* conservation area of 0.15 ha of *Alnus jorullensis*. Also here a nursery has been established producing *Alnus jorullensis* for planting on terrace risers thus 1) fixing nitrogen to benefit agricultural crops, 2) providing firewood and 3) providing soil protection. These activities are considered an example of an optimal strategy of *in situ* conservation. The small natural, remnant stand of *Alnus*, belonging to the community, was being over exploited for wood, posts and poles, and its management was being totally neglected. By demonstrating to the campesinos that food production – a day-to-day concern to them – could be greatly improved by using material originating in the *Alnus* stand, planting *Alnus* seedlings along the contour lines of agricultural fields, the value of this earlier-neglected stand suddenly increased by many orders of magnitude. At the same time, demonstrations were made to show that although the stand could, and should, continue to be used also for the provision of wood and wood products for day-to-day use by the community, management (including ensuring regeneration by careful felling and extraction) visibly improved it and was apt to ensure the availability of such material also in the future. The locals have come to realize the practical value of this stand in their daily life, as well as being aware of the fact that the stand is also of importance to 'the Government' (translating into the value of conserving the genetic resources of this specific stand, as one component of a network of *in situ* conservation areas of *Alnus jorullensis*).

"The management of natural, remnant *Alnus* stands and growing and utilization of seedlings with materials originating in such stands has already spontaneously spread to other, neighbouring communities" (FAO, 1990b).

[6]Urban forestry has been defined as "a specialized branch of forestry that has as its objectives the cultivation and management of trees for their present and potential contribution to the physiological, sociological and economic well-being of urban society. In its broadest sense, urban forestry embraces a multimanagerial system that includes municipal watersheds, wildlife habitats, outdoor recreation opportunities, landscape design, recycling of municipal wastes, tree care in general and the production of wood fibre as a raw material" (Kuchelmeister and Braatz, 1993).

[7]*Cinara cupressi* attacks the tree species *Cupressus lusitanica*. *Heterospylla cubana* attacks the tree species *Leucaena leucocephala*. *Rhyacionia buoliana* occurs on pines in its native European range but has recently been discovered attacking *Pinus radiata* plantations in Chile; *Sirex noctilio*, native to southern Europe, has attacked *P. radiata* plantations in New Zealand and is now attacking *P. radiata*, *P. taeda* and other introduced pines in Argentina, Brazil and Uruguay.

[8]The experience of the Aracruz Florestal, responsible for the plantations of Aracruz Celulose, a Brazilian paper pulp company, provides an example of the potential of plantations for the production of industrial roundwood. The annual increment of the first plantations of eucalypts was 28 m^3/ha per year, but natural hybrids, mainly between Eucalyptus grandis and *E. urophylla*, increased this to 45 m^3/ha and synthetic hybrids of *E. grandis x E. urophylla* now yield 70 m^3/ha annually (FAO, 1993j).

CHAPTER 6

Laws, institutions and people

Most countries have regulations and an administration
for the forestry sector, within which forest management is implemented.
But often the laws are out of date and are neither enforced nor enforceable.
Institutions are weak and are not adapted to their tasks.
Change is needed to create an enabling, incentive framework that will
give the people of every interest group a voice in the decision–making
process and a share in the benefits of forest management.

Regulatory framework

A sound and effectively implemented system of land-use planning under which areas of forest are demarcated and set aside as permanent forest estate is an essential prerequisite of all sustainable forest management systems.

Control over access is particularly important if the areas set aside are to be properly managed. This is a major difficulty in most countries in the developing world. Even formally classified forests are in effect unprotected in many areas and are subject to encroachment, illegal logging and charcoal-making and other destructive pressures – let alone the large areas of undemarcated forest that have no formal status.

In the past, forests in former colonies were regularly demarcated by colonial administrations but almost invariably in a manner that ignored local people's rights and indigenous management systems and not always on sound land-use principles. In many areas, people were forcibly excluded from areas in which they had lived and farmed and over which they had exercised traditional rights for generations. Although these reserves prevented encroachment or illegal logging in many areas, this was usually at the price of considerable local ill-feeling towards the forest-service officers who policed them.

Some countries on gaining independence immediately handed over some of their official forest lands to farmers. In many countries, however, the reserves still exist and continue to protect the forests. But the legacy of antagonism between forestry officials and the local community tends to persist and remains a serious obstacle to forest management.

Effective demarcation of forest areas is still a difficult and complex issue. The fact that the majority of tropical forests are formally owned by the state does not lessen the need for sensitivity in designating particular areas for management if a repetition of the colonial experience is to be avoided. Demarcation of forest lands without popular consultation will almost inevitably fail in its purpose, because regulation in itself will never be successful in controlling access over large areas. Similarly, arbitrary demarcation that bears no relation to the ecology or quality of the land can also lead to problems. It is essential that the designation of forest lands for particular purposes be based on sound land-use planning principles.

The enforcement of forest legislation emphasizes its restrictive, "policing" function, while at the same time restrictions on tenure and uses of trees act as a disincentive.

In some countries the basic

problem is not that forests lack legal protection, but that the legislation covering the allocation of land titles and fiscal policies actively encourage deforestation. In these countries, people clearing forest can claim title to the cleared land. This is a system that encourages destruction of forests; the felled trees are often not even used or sold for timber but are simply burned. The process is graphically illustrated by a number of cases in Latin America where the forest services have had to give up awarding logging concessions because the forests never lasted long enough for the concessionaires to harvest the timber. The reason was that agriculture departments were encouraging settlers to clear forests as a way of acquiring formal title to the land. At a time when the area of forest relative to the population seemed limitless and expansion of the agricultural frontier was seen as a primary national concern, such regulations made sense. At a time when reducing the rate of deforestation is an urgent concern, they are counter-productive.

Complete prohibition of forest clearing in agricultural frontier areas is unrealistic. It is nonetheless possible to reframe legislation so that rather than encouraging waste it would promote a more rational use of resources. Thus, forest land should be surveyed in advance to assess its suitability for conversion to sustainable agriculture. Areas that are designated as suitable for conversion should then be released in a properly planned and controlled manner. Arrangements should also be made to ensure that the trees on land to be cleared are utilized instead of being burnt.

Forestry services and public administrations

Forestry services and public administrations, as the main repositories of formally recorded forestry knowledge and experience, have an important role in developing strategies for the management and protection of the world's forests. Unfortunately, many forestry services, particularly in the developing world, are ill-equipped for even their technical tasks, let alone for taking a wider role in the development of the national land-use and rural development strategies of which sustainable forest management is ideally an integral part. Without substantial amounts of additional funding from forest royalties or other government sources, it is unlikely that a high proportion of developing country forestry services and institutes will be able to meet the challenges facing them.

Forestry services are no longer required to fulfil technical functions only. Their role has expanded to cover social issues, interaction with other sectors and the integration of the forestry outputs within national development plans. The question is now whether forestry services have the means and the skills commensurate with the magnitude and complexity of the task they face.

Strengthening and reform of forest services are essential if they are to be able to make the contribution required of them in the future. The most basic requirement is that forestry should be given a greater allocation of government resources. With many other urgent pressures on the governments of most developing countries, it is difficult to see where these resources are to be obtained. But a reformed royalty system, in which all or most of the proceeds would be retained by the forestry sector, could increase the total funding available. This would also provide a greater incentive for forest services to be more diligent in the collection of fees and royalties than they are at present, although the forest service itself would have to be closely monitored by an independent body to ensure that this power not be abused.

Forest policies and forest laws have to be continually revised and updated if they are to remain relevant. The most important concept to introduce into present-day policies and laws is the need to secure people's participation in forest resource management – including participation in the policy-making process itself – while ensuring sustainability of the resource through safeguards on control. Unfortunately many countries, even those with long-established forest services, have not carried out such revisions. Another weakness in the forest services of some countries is the lack of planning capability, and often even of the reliable figures that are the raw material of the planning process. Forest services must develop channels for communication. Extension agents must provide fora to help foresters learn from farmers (or other groups), to help farmers learn from the forest service and to help farmers share information with one another. The means are required for information to flow upwards as well as downwards, so that managers and policy-makers are informed about the realities of the rural situation and the success of their programmes.

Privatization

The trend of the past decade to divest the state of many of its functions and hand them over to the private sector is unlikely to end in the near future. The reduced bureaucracy, greater competition, greater accountability to the public and other benefits achieved by the privatization of many public bodies keep its appeal high. In the forestry sector, the European experience, especially in Scandinavia, shows that private management of forests can work well provided there are adequate incentives and a framework of rules which are respected.

But it also has to be borne in mind that privatization in itself does not necessarily lead to the maximum public welfare, especially in the environmental area. The World Bank in its 1991 forest policy paper bluntly stated that "the free interplay of market forces will not bring about socially desired outcomes" (World Bank, 1991). *The Economist* magazine said in an editorial on the environment:

> *The market is unlikely to help. Even in an individual country it will rarely deliver what is best for the environment. The costs to individuals and companies of polluting or pillaging the environment will be lower than the costs their activities impose on the rest of society. National environmental policies therefore need government to step in and ensure that polluters carry the costs they would otherwise dump on their fellow citizens (Economist, 1992).*

A danger of the privatization of forests is that it would shift the planning horizon and the discount rate applied to investments to benefit individuals or commercial companies rather than society as a whole. As a result, investments that underpin public welfare and that have traditionally been the responsibility of governments, for example in forest or watershed management, may not be made. Another danger is that when land is privatized whole groups of people may be cut off from access to certain resources upon which they depend. It is often the poorest who depend most heavily on communal land for their livelihood.

Privatization must therefore be accompanied by consultation and by the establishment of a regulatory framework that ensures that the decisions taken by private companies are in accord with the overall welfare of society. The institutions necessary for interpreting and implementing these regulations must also have the necessary funding, competence, experience and stability. Failure to protect the forests effectively against the inherently short-term forces of the market could clearly have disastrous results. This is the main reason why in many countries privatization only affects forestry operations and forest industries. In countries as diverse as New Zealand and China public lands on which forests grow remain the property of the state in order to ensure long-term continuity in management. There is thus every reason to proceed slowly and cautiously towards their privatization.

Training

Training is required for all people involved with the management of forests, at every level, whether in the public service, in the private sector or in rural cooperatives.

Existing forest-service staff will have to change attitudes, learn modern management techniques and learn to respond to the new demands upon them. It is no longer possible to retreat to a closed technical world inside the forest boundaries. Terms such as social forestry, participatory forestry and community forestry may be irritatingly vague to many who have been trained in the traditional way, but they represent important new dimensions in practising forestry. In-service training is required for staff in every forestry service, but at present is often lacking, especially in many developing countries. Training will have to have particular emphasis on working with and learning from local communities, women and farmers. Confrontation was never a good idea and in today's increasingly crowded world it is futile and self-defeating. Rural development in the broad sense is the only means that will reduce and eventually eliminate

Training at all levels, in both the public and private sectors, is vital if people are to learn to respond to the new demands of forest management.

the pressures causing deforestation. Forestry practice will involve collaboration with and participation of rural people and other organizations, both public and private, and forestry services will have to redefine their role to comprehend planning, developing an enabling environment, providing technical support and monitoring.

At the same time, it is essential to retain the professional core of forestry with its roots in accumulated knowledge and experience. There is no point in trying to turn trainee foresters into sociologists, anthropologists, agriculturalists, ecologists and extension agents all rolled into one. They must, however, be made aware of the relevance of the various disciplines, seeing their own role in a much wider context than they have traditionally done and learning how to work in multidisciplinary teams.

Research

Research is a critical area of concern. Natural forest management involves dealing with the whole forest ecosystem, as well as the individual species, over prolonged periods. It needs to be supported by detailed research if the techniques used are to be refined and developed.

The long-term nature of forestry research requires a corresponding stability in the staffing and funding of the institutions involved. It is not only the growth and change of complex forest ecosystems that take place over long periods of time; training and development of scientific expertise are also long-term processes. The academic training of young scientists takes six or seven years, and it can be ten years or more before they are capable of original independent work in their specialization.

There is no lack of forestry research topics on which work is urgently required. They range from the fundamental to the applied, from the technical to the social, from the economic to those concerned with policy. They include detailed studies of forest ecosystems, the silviculture and biology of tropical forest species, techniques for the conservation of genetic resources and biodiversity, improvements in management and harvesting systems, market development of currently non-commercial species, protection against diseases and pests, improvements in the output of non-wood products, increases in the understanding of relationships between forests and local communities and between forests and the environment, and identification and quantification of the benefits of forests.

The range of topics extends into

agriculture, since without stable and productive agriculture the survival of the forests will remain at risk. Researchers must learn from the farmers, not just of the challenges that they face but also of the solutions they have already identified. Furthermore, specialists now work with farmers to establish the farmers' conception of the ideal tree (or other plant), known as the ideotype, and then to identify the available genetic material or the role of mechanical manipulation (pruning, etc.) to use to approach the ideotype.

Many of these problems can be tackled by joint research programmes among institutions in different countries. Such links are being strengthened through new international institutions devoted to forestry research, which are discussed in Chapter 7. Nonetheless, local research must always be stressed, as must the transfer of research findings to the field, which remains a weakness still to be adequately addressed.

Involving rural people

Most forests in the developing world are on lands on which indigenous groups and other rural communities depend. It is essential that these people be involved in any proposals for the management of the forests.

It is frequently assumed that rural people do not know the value of environmental protection. This is condescending and almost invariably wrong – what these people need from outsiders is not exhortation or advice, but practical help in doing what they are well aware needs to be done.

Schemes that depend on setting aside forest or farming land for environmental or social reasons will not work satisfactorily unless local people feel that their legitimate demands are being met and that they are obtaining a fair share of the benefits. This will often require a much more open attitude on the part of forestry services, government officials and donor agencies.

Often there has been an assumption that rural people do not know the value of environmental protection. This view is condescending and almost invariably wrong. Rural people are only too aware of forest values and often have the solutions to the management problems associated with protecting them. The reason these are not applied is that there are other pressures, concerns and constraints. What rural people most often need from outsiders is not exhortation or advice but help in doing what they are well aware needs to be done.

It also has to be borne in mind that growing or preserving trees is seldom the first priority of rural people. An experience of a development project in the Sudan illustrates the point. After an entertaining puppet show that advised the community to plant trees, one of the community leaders spoke up: "Most of us are landless here. If you can bring us land we will

With increasing populations, traditional landownership systems are breaking down.

first grow food for our children and then we would be glad to plant your trees for you." In other words, in this case food security was the pivotal issue around which virtually everything else turned.

As populations increase and the pressure on the available land becomes heavier, many countries are finding that traditional common ownership systems are breaking down. People are enclosing and privatizing their lands as a means of protecting their own interests. This provides an additional incentive to invest in long-term land improvement measures. In some of the more densely populated areas of the Kenya highlands, for example, surveys have found increasing numbers of trees being planted on individual farms as the common resources decline or are privatized.

Private ownership of land, under certain conditions, may bring security to families as well as improved agricultural management and increased tree growing. But privatizing land may not always be the optimum solution, especially in the drier areas where the land is sequentially used by farmers and herders at different times of the year. In areas where common land management systems are breaking down, rather than assuming that a private tenure system will automatically bring an improvement, it is better to find out why the old system has failed and whether it can be restored.

Sometimes the best solution may simply be to trust the local population. One of the boldest moves of this kind to date has been made in Colombia, where the government has returned legal control of over 18 million hectares of land in the Amazon area to some 70 000 native people. There is, of course, no guarantee that forests will be managed wisely or sustainably in the hands of local people. In Fiji and Papua New Guinea, the local groups or clans who owned the forest lands were misinformed of the true intentions of loggers, and their forests have suffered the same overexploitation and destruction as in other countries. But in Colombia there is at least a reasonable chance that the results will be more equitable socially and more favourable environmentally than they would be if control were to rest entirely with cattle ranchers or logging companies.

Another example of cooperation with local people, which is already having major success, is in West Bengal. By the early 1980s, huge areas of the natural sal forests had been reduced to degraded scrub by overgrazing, fires and the cutting of trees for poles and fuelwood. The potential for regrowth, however, remained high as the trees coppice easily. After the failure of many attempts to realize this potential by policing the forests and arresting fuelwood cutters and dealers, the forest services began to work with local communities, providing them with a share in the proceeds from regeneration of the forests. Now, local communities and the forest services are jointly managing over 250 000 ha of luxuriantly growing forest which can be utilized for timber, poles, fodder and other outputs.

The role of non-governmental organizations

Over the past decade, non-governmental organizations (NGOs) have become increasingly involved in forestry activities. They vary greatly in size, competence and objectives. Some are large international organizations with budgets many times greater than those of most national forestry services, while at the other end of the scale are small local voluntary groups. Many of the latter represent the village people. Their activities range from international discussions with government representatives and donor agencies to practical involvement in the problems of villagers living at subsistence level.

Most NGOs have a very different approach to their work than forest departments and other government agencies. Their goals are usually more limited, and because many have no statutory powers they must rely entirely on persuasion of the local people with whom they are working. As a result, they necessarily have to develop effective methods of communication with local communities. They also tend to be smaller, less bureaucratic and more flexible in their approach than official agencies, though this is decreasingly true of some of the larger international NGOs.

NGOs have their limitations. Some become involved in issues beyond their capacity and promise more than they can deliver. Some become self-righteous and intransigent, focused more on their image and fund-raising needs than on addressing the problems being faced at the local level. Some may have a short lifetime, may be set up purely opportunistically or may lack the resources or capacity for long-term projects. To avoid raising expectations that cannot be met, NGOs must be open and must accept scrutiny of their performance and the need for accountability.

To date, most NGO involvement has been concerned with the social and environmental impacts of forest utilization rather than the practicalities of production forestry. But in some places this is changing. In the Philippines, the government is collaborating with NGOs in the Community Forestry Programme which encourages people living near forests to carry out inventories and manage the forests for sustainable timber production.

Because, in general, NGOs do not start with a concern for timber production, which is at the heart of forestry tradition and training, they tend to take a much broader approach to forest-related issues. They are prepared to question many of the assumptions on which forestry policies are based. They are also able to fill in gaps in policies, providing support and advice for local communities which official agencies have been unwilling or unable to give.

NGOs can be particularly useful in representing women. Because of gender-based divisions of labour, women have specific needs and interests, especially in relation to the provision of food and fuel for their families, that are often not adequately taken into account in forestry projects. Social customs may also deny them a hearing in policy discussions even about projects where their attitude to what is being proposed is the most decisive factor.

Many social forestry programmes, for example, have failed because their planners did not consult the women who were expected to benefit from them but who proved unwilling or unable to carry them out. Programmes that involve extra work for women, for example, may be impossible to carry out unless there is a reduction of their workload in other areas. By involving and consulting NGOs that represent women's interests, such problems and design mistakes can be largely avoided.

The involvement of an appropriate organization representing local people may well be the key to the success of a community-oriented forestry programme. But collaboration with state agencies may create problems for the organization, such as the costs and frustrations of dealing with the bureaucracy and a restriction of the power to act independently. Working with state agencies can also reduce credibility, with the possibility that the NGO will be seen as an agent for the implementation of locally unpopular projects. These dangers can only be avoided if organizations that represent local people are genuinely allowed to be part of the decision-making process and are publicly seen to be doing so.

NGOs are here to stay. They have demonstrated that they can make major contributions not just to the local implementation of forest policies but to their initial design and development. But simply involving an NGO is unlikely to bring any benefits unless there is a genuine commitment to collaborative working.

CHAPTER 7

The international dimension

A number of important international initiatives have already
been taken to slow the present rate of depletion of the world's forests and
eventually to stabilize their area. There is also significant public
interest and involvement in proposals such as debt-for-nature swaps and the
steps taken in some industrialized countries for consumer
boycotts of tropical timber originating from land that is
not under sustainable management.

Official development assistance (ODA) from donor countries, development banks and United Nations organizations to the forestry sector in 1990 was US$1 353.5 million, or an increase of US$260 million (24 percent) over the figure for 1988 (FAO, 1991b). The value of ODA to the forestry sector in 1988 represented 9.2 percent of the ODA to the agriculture sector and 2.8 percent of the ODA for all sectors. About two-thirds of the ODA to the forestry sector came from donor countries, with the balance approximately equally divided between the development banks and UN organizations. The main fields of expenditure of the 1990 ODA were: forestry in land use (35 percent), forest-based industrial development (28 percent) and institutional strengthening (19 percent). The proportions of the allocations among the different fields of expenditure remained much the same as in 1988, except that the share devoted to fuelwood fell from 17 percent in 1988 to 6 percent in 1990; the reasons for this reduction are not clear.

International organizations and agreements

Support for developing country forestry projects is provided by a variety of national and international agencies. Increasingly, sustainable forest management is featuring in policy documents as well as in loan and grant agreements.

The World Bank is the largest single multilateral lender; up to 1991 it had provided nearly US$2 500 million in loans to a total of 94 forestry projects. In June 1991 it produced a new forest policy document (World Bank, 1991), its first since 1978. This identified the following as the two main challenges to be met: the prevention of excessive deforestation, especially in the tropical moist forests, and action to meet increasing demands for forest products and services for the rural poor in developing countries through tree planting and the management of existing forest resources. The policies proposed for forest protection to check deforestation include: poverty alleviation, the allocation of land to forestry purposes, corrections to private incentives that encourage timber "mining" and the clearing of forest land, and a controlled increase in public investment. Policies to meet the basic needs for forest products and services arising from forests would include reductions in demand and increases in supply. Other requirements that were recognized were the strengthening of forest institutions and the involvement of the international community in institution building, financing and international trade reforms.

The World Bank has been attacked for promoting environmentally and socially damaging forestry projects because of its emphasis on the economic rates of return on its loans. In its new policy document the bank responded to the criticisms by pledging that future project evaluations would distinguish between projects that are environmentally protective or small-farmer oriented and those that are purely commercial. It also promised that its lending in the forestry sector would be subject to commitments by governments to sustainable and conservation-oriented forestry. One condition laid down in the document is that the World Bank "will not under any circumstances finance commercial logging in primary tropical moist forests". While this policy may reflect the concerns of many environmental groups for such ecosystems, it will not contribute to their conservation.

Since its foundation in 1945, FAO, which is the main UN agency mandated to deal with forestry, has played a major role across the whole spectrum of forestry activities throughout the world. It cooperates with member countries in rural development, policy analysis, institutional strengthening, plantation forestry, community forestry, participatory forestry, sustainable forest management, education, training and the production of scientific information and global statistics on forestry. It provides a neutral forum for discussion on global and regional forestry matters, through, for instance, the biennial meetings of the Committee on Forestry (COFO) and the Committee on Forest Development in the Tropics and the regular meetings of the various regional forestry commissions. The Organization has permanent missions in a large number of member countries which serve to link FAO's expertise with that of national governments; it is also the implementing agency of many donor-funded projects.

In 1985 FAO, the United Nations Development Programme (UNDP), the World Bank and the World Resources Institute (WRI), a Washington-based privately funded research organization, co-sponsored the Tropical Forestry Action Plan (TFAP), renamed in 1991 the Tropical Forests Action Programme. TFAP was the result of a number of tropical forestry initiatives undertaken in the early and middle 1980s. TFAP emphasizes management and planning within the context of overall land-use development as well as the need to involve rural people; it broke new ground in explicitly recognizing the importance of NGOs and allocating them a pivotal position in linking local communities and national governments.

TFAP provides a framework for national programmes for sustainable forest utilization and for harmonizing and strengthening international donor cooperation. FAO supports the efforts of participating countries through its coordinating unit.

By May 1993 TFAP had the formal participation of 90 countries[1] and had helped 33 countries draw up national forestry action plans that were being implemented, generally with donor support; a further five regional plans were under preparation and one is being implemented. It had also helped to facilitate a 16 percent annual increase in development assistance funds to the forestry sector. Within the framework of TFAP this amounted to a total of US$1 300 million in 1990 and US$2 030 million in 1992.

During its first years of operation TFAP came under some criticism. WRI monitored NGO participation between 1988 and 1990 and found that, in practice, it was far less than had been expected, with some governments being extremely reluctant to concede NGOs any meaningful involvement. The study also showed that some of the participating countries put much emphasis on timber production as compared to conservation and rural development. Some tropical governments were disappointed, feeling that the international community had been ungenerous in its provision of assistance.

The findings led to a shift from donor coordination and project activity to a longer-term programme emphasizing policy development aimed at the conservation and sustainable development of forests. Country concerns and activities were to be given greater attention (country-driven emphasis) while plans and programmes were to take account of the interests of the related sectors (interdisciplinarity), with greater involvement of the people and their organizations. Some NGOs still remain sceptical, alleging that donor funding remains based upon previous policies rather than upon the quality of the action plans or the priorities of the recipient nations. Supporters of TFAP maintain, however, that time is needed for its full benefits to be realized, and that for all its flaws it is uniquely valuable for facilitating cooperative efforts between tropical countries and the international community.

The International Union of Forestry Research Organizations

(IUFRO), founded in 1892, links forestry research institutes into a global network of research units (interest groups). Its Special Programme for Developing Countries (SPDC), set up in 1983, supports training and self-teaching programmes and information services and promotes interagency contacts.

Whether the role of SPDC will change with the establishment of the Centre for International Forestry Research (CIFOR) remains to be seen. CIFOR, based in Bogor, Indonesia, was established in February 1993 as a member of the Consultative Group on International Agricultural Research (CGIAR). This new organization has a global mandate but will operate in a decentralized fashion through continental coordinators, networks and collaborative research. CIFOR's focus will be on conservation and the improved productivity of forest ecosystems, with programmes in natural forests, open woodlands, plantations and woodlots, and degraded lands. It will include an important component for the strengthening of national forest research institutions.

The International Council for Research in Agroforestry (ICRAF), which has been based in Nairobi since its establishment in 1978, has also recently joined the CGIAR. It has a mandate for global agroforestry research and will collaborate closely with CIFOR. It is currently promoting "Alternatives to Slash and Burn", a programme carried out in collaboration with research institutions in Southeast Asia, Africa and Latin America.

The International Board for Plant Genetic Resources (IBPGR), established as an FAO body under mandate from the UN Conference on the Human Environment (Stockholm, 1972), will shortly become a fully autonomous and independently administered centre within the CGIAR system. IBPGR is preparing proposals for its involvement in the conservation of biodiversity and the germplasm of forestry species.

The International Tropical Timber Organization (ITTO), established in Yokohama, Japan, in 1985, is made up of the world's major tropical timber exporting and importing nations. It supports various programmes concerned with timber harvesting, forest management, reforestation, wood utilization and marketing as well as providing a forum for the international timber trade. ITTO published guidelines on sustainable natural forest management in 1990, on planted tropical forests in 1991 and on biodiversity conservation in 1992. These are intended as broad guides to be followed but to be modified to suit the situations and conditions in individual countries.

ITTO has established a strategy of seeking to encourage member countries to move towards sustainable management of tropical forests. Part of this strategy involves the target that all exports of tropical timber will come from sustainably managed resources by the year 2000. Producer member countries have committed themselves to achieving this, but have stressed that their success will depend on substantial financial and technical assistance from consumer countries.

At a regional level, the Mediterranean environment has been a matter of growing international concern. The forests of the countries to the north of the Mediterranean Sea have been mainly under formal management, while those of the tropical and subtropical countries to the south have been included in TFAP. The countries of the Mediterranean basin have drawn attention to this apparent neglect of a rich and unique forest zone and have proposed an international support programme.

In conformity with the Paris Declaration which was issued after the Tenth World Forestry Congress in 1991, the Forestry Department of FAO, acting as the Secretariat of Silva Mediterranea, drafted a Mediterranean Forest Action Programme. This programme, known as MED-FAP, has been drawn up and is due to be launched in 1993. It aims to help each country carry out a rapid assessment of the state of its forests, identify its priorities, draw up short- and medium-term action plans and obtain funding for their implementation. Such activities are also in conformity with the recommendations of the United Nations Conference on Environment and Development (UNCED), which urged every country to establish its own Forest National Plan, with the support of the international community if necessary.

The World Wide Fund for Nature (WWF) is concerned with fund-raising and increasing awareness; it has interests in species and protected area conservation and in forest management. The World Conservation Union, formerly known as the International Union for the Conservation of Nature and Natural Resources (IUCN), has a mandate to promote action and to seek support for projects related to environmental protection and resource management. These two NGOs are international in the sense that their activities are not directed by members in individual countries.

The United Nations Conference on Environment and Development

UNCED, which was held in Rio de Janeiro in the summer of 1992, devoted a considerable amount of attention to the question of the world's forests (see Lanly, 1992). The Rio Declaration, which was unanimously adopted at the end of the conference by the 102 heads of state and government attending, proclaimed 27 general principles to guide states and people in a "new and equitable global partnership" in matters of the environment and of development and in related policies and programmes. Several of these principles are concerned with sustainability, such as those calling for environmental protection as an integral part of development (Principle 4), reduction and elimination of unsustainable patterns of production and consumption (Principle 8), adoption of effective and appropriate environmental legislation (Principle 11) and environmental impact assessments at the national level (Principle 7). Principle 2 stated that sovereign countries have the right to exploit their own resources according to their own environmental and developmental policies. The conference proposed that the Rio Declaration should be further elaborated and presented as an Earth

Charter on the 50th anniversary of the UN in 1995.

At the same time, the conference drew up a "non-legally-binding authoritative statement of principles for a global consensus on the management, conservation and sustainable development of all types of forests" (the "forest principles"). The Preamble, which begins by stating that "the subject of forests is related to the entire range of environmental and development issues and opportunities, including the right to socio-economic development on a sustainable basis" (Paragraph a), also states that "forests are essential to economic development and the maintenance of all forms of life" (Paragraph g). The Preamble draws attention to the multiple and complementary functions and uses of forests and to the need for a holistic and balanced view of the issues and opportunities for their conservation and development. It states that the principles apply to "all types of forests, both natural and planted, in all geographic regions and climatic zones" (Paragraph e).

Countries have committed themselves to the prompt implementation of the forest principles and to "consider[ing] the need for and the feasibility of all kinds of appropriate internationally agreed arrangements to promote international cooperation on forest management, conservation and sustainable development of all types of forests".

The provisions of the forest principles are wide ranging and cover all aspects of the management, conservation and sustainable development of forests all over the world. The need for sustainability is stressed throughout. For instance, Article 2b asserts that "forest resources and forest lands should be sustainably managed to meet the social, economic, ecological, cultural and spiritual human needs of present and future generations" and Article 3a that "national policies and strategies should provide a framework... for the management, conservation and sustainable development of forests and forest lands".

The forest principles define the sovereign right of countries over their forest resources (Articles 1a and 2a). Participation in forest conservation and development is recommended for some specific sections of society, such as women (Articles 2d and 5b), indigenous people and forest dwellers (Articles 2d and 5a) and industries, labour, NGOs and individuals (Article 2d). The need for institutional strengthening is stressed in general (Article 3a), in relation to

research and training (Article 12b) and in relation to the use of the capacities and knowledge of indigenous and local communities (Article 12d). The protection of ecologically viable representative examples of primary or old-growth forest as well as forests valued for their cultural, spiritual, historical, religious and other values is emphasized (Article 8f). Article 15 calls for the removal or avoidance of unilateral measures, which are incompatible with international obligations, to restrict international trade in timber or other forest products in order to attain long-term, sustainable forest management.

A final agreement, "Combating deforestation", which appeared under Agenda 21, Chapter 11, was adopted by the plenary session of the conference on 14 June 1992. This covered four programme areas:
• sustaining the multiple roles and functions of all types of forests, forest lands and woodlands;
• enhancing the protection, sustainable management and conservation of all forests, and the greening of degraded areas, through forest rehabilitation, forestation, reforestation and other rehabilitative measures;
• promoting efficient utilization and assessment to recover the full valuation of the goods and services

provided by forests, forest lands and woodlands;

• establishing and/or strengthening capacities for the planning assessment and systematic observations of forestry and related programmes, projects and activities including commercial trade and processes.

Originally there was a fifth programme area covering international and regional cooperation, but its components were finally distributed among the other four programme areas. TFAP is included under the second programme area, that concerned with enhancing the protection and sustainable management of all forests, which is linked to an appeal to all countries to formulate and implement national forestry programmes or plans.

The UNCED Secretariat estimated the total annual cost of these programmes at about US$30 000 million per year, of which about $5 500 million would come from the international community on grant or concessional terms. Discussion of how and from where such funding should come was not on the UNCED agenda, though a number of countries and organizations used the occasion to promise or pledge financial support.

The fact that the issue of sustainable management of all forests is now firmly on the international agenda is

encouraging. But as the haggling and arguing at the conference showed, there is a long way to go before the ringing declarations are translated into the commitments in terms of both policy and finance that are required if effective action is to take place.

UNCED also agreed upon a framework convention on climate change, which was signed by 150 countries. This will enter into force after it has been officially ratified by 50 national governments from among the signatories. The convention mainly commits countries to reductions in greenhouse gas emissions and includes proposals for helping developing countries. No targets on emissions or deadlines are, however, included in the convention.

A framework convention on biodiversity agreed upon at the conference was signed by 154 countries. This will come into force after national ratification by 30 countries. Its objectives are "the conservation of biological diversity, the sustainable use of its components and the fair and equitable sharing of the benefits arising out of the utilization of genetic resources and by the appropriate technology".

The value of the principles enshrined in the Rio Declaration and in the non-legally-binding forest principles and the efficacy of the conventions on climate change and on biodiversity will depend to a large extent on the political will and commitment of the governments and policy-makers to apply their tenets. The body charged with monitoring the implementation of the agreements reached at UNCED is the UN Commission on Sustainable Development (CSD), an intergovernmental forum with participation at senior (including ministerial) level.[2]

Debt for-nature swaps

Many developing countries are heavily burdened with foreign currency debts to commercial banks, to the governments of industrialized countries and to international lending agencies. For some of the poorest countries the possibility of paying off these debts within decades, if ever, is extremely low. Banks and lending agencies, however, remain reluctant to cancel such debts since they consider that it would lessen the chances of being repaid by those countries that are better able to afford the repayments.

Debt-for-nature swaps (DFNS) represent an attempt to benefit the environment by capitalizing on banks' willingness to discount (or sell at less than face value) some of their long-standing debts, to accept an immediate reduced payment rather than waiting for full payment, which

may never come. The swap is organized by a third party, either an environmental NGO or the government of an industrialized country, which buys the debt at a reduced rate from the lender and agrees to cancel it provided the debtor country makes a specified investment in an environmental project such as protecting an area of tropical forest of particular ecological significance. In this way, the developing country rids itself of the debt without having to repay any foreign currency but assumes responsibility to undertake certain environmental activities

There are various instruments for debt swaps (Bramble, 1987), which include:
• debt swaps: the exchange of one country's debt for that of another;
• debt-for-equity conversions: the conversion of a hard-currency debt into an investment in the developing country;
• debt-for-commodity conversions: the conversion of debt into an export commodity, which can be sold to meet the discounted value of the debt;
• factoring companies: the purchase, by a company established for the purpose, of problem loans from banks in exchange for dividends paid by the factoring company from trading or investing the debt;

• bonds: a variation of long-term bonds, "exit bonds" are denominated in hard currencies but can only be traded for local currency before final maturity.

By the end of 1992 a total of 24 debt-for-nature swaps had been arranged with a face value of over US$122 million, generating conservation funds of over US$75 million but in fact costing just over US$23 million (see Annex 1). The overall impact is small in relation to the total debt of the developing world and the rate at which deforestation is taking place. There are also problems in ensuring that the conservation programmes are realistic and capable of being implemented and maintained on a sustainable basis. An initiative in Bolivia, for example, ran into trouble because local indigenous people as well as loggers who had been awarded concessions laid claim to the land that was to be managed.

Objections have been raised to the rationale of debt-for-nature swaps. They have been criticized as a form of "eco-colonialism", under which industrialized countries would set the priorities of developing countries. It is possible that the system would work in countries that are badly mismanaging their economies, but not in others that are displaying economic discipline. The swaps may add to inflationary pressures. As a result of the attraction of the economic and ecological benefits of debt-for-nature swaps, the needs of the people living in the area may be overlooked. It is nevertheless generally accepted that prudently chosen and well planned debt-for-nature swaps are a viable, if rather complicated, measure which can make a contribution to forest conservation.

Genetic materials and medicines from the forests

An area of controversy in which there is an increasingly evident need for an international agreement is the search by companies from industrialized countries for potentially useful drugs and genetic materials from tropical forests. A successful discovery can be extremely lucrative for the companies involved.

The large-scale harvesting of such drugs from the wild is generally not the objective of these investigations,

although in the testing phase large quantities of the raw material are likely to be required. The hope of the companies is that their researchers, by examining the properties and roles of various plants or other forms of life in forest ecosystems, will identify chemical compounds that can be used for human benefit. Normally, once such a compound has been discovered and fully tested, it might be synthesized (and protected by patents); the desired genes might be transferred into another organism; or the plant from which it was derived might be grown under cultivation.

It is hoped that by giving the countries in which the original discoveries are made a share in the profits arising from the use of the genes they will be given an incentive to conserve the ecosystems from which the genes are derived. Contracts could, for example, be based upon exploration fees and a royalty on any drugs successfully developed and marketed, but complex arrangements will be required to monitor sales and to give the company continued (but not sole) access to the genetic resources.

The best known example of an agreement between an institution in a developing country and a private pharmaceutical company for exploration of the potential of natural species for commercial development while at the same time promoting conservation of those species is that between INBio (the National Biodiversity Institute of Costa Rica) and Merck & Co., a large American pharmaceutical firm. Under this agreement, which was reached in 1991, Merck pays INBio to sample and process, for drug screening purposes, the range of plants, insects and other organisms naturally occurring in Costa Rica. Merck will also pay INBio royalties arising from any drug that is commercially developed from the screening. INBio in turn pays a proportion of the budget for exploration and screening; it also pays a proportion of any eventual royalties to the Costa Rican government for conservation purposes (World Resources Institute, 1993).

The issue of recompensing governments and peoples for the natural genetic resources of their countries was addressed at UNCED within the context of the framework convention on biodiversity, which called for sharing "in a fair and equitable way the results of research and development and the benefits arising from the commercial and other utilization of genetic resources... upon mutually agreed terms".

But the exploration of biodiversity will not necessarily lead to investment in the conservation of forest ecosystems, nor to rich and immediate rewards to the countries where the plants or animals with the desired genes are indigenous, nor even to a share of the rewards for the people using and presently conserving the resource. That there are potential returns from the exploration of the biodiversity of forest ecosystems is undeniable, but if such exploration is to contribute to sustainable forest management it will have to be controlled and managed in such a way that the returns benefit the indigenous people who have an interest in conserving the resource (Reid *et al.*, 1993).

Consumer boycotts

Consumer boycotts of tropical timber have been promoted by a number of environmental NGOs in the industrialized countries in recent years. These have had an impact in some countries. About 450 city councils in Germany, for example, have banned the use of tropical timbers and over 90 percent of Dutch local councils have done the same. In the United States, the states of Arizona and New York have banned the use of tropical timbers (excluding tropical plywood in New York) in public construction projects, as have several cities (*Global Environmental Change Report*, 1991).

Certain tree species have now been listed in the Convention on International Trade in Endangered Species of Wild Fauna and Flora (CITES). CITES is an international convention which governs trade in named endangered species by voluntary agreement among its member countries (99 in 1989). These species are either threatened by extinction or if not now threatened may become so unless their trade is subject to strict regulation.[3]

The objectives sought from bans and boycotts vary, and hence so does their likelihood of being effective. Some seek to encourage consumers to stop buying all tropical timbers in order to maintain species diversity and ecological systems; others are intended to promote sustained yield management. Such conflicting objectives make it difficult to convey a clear message to tropical timber producers and reduce the chances of achieving useful results.

Even if boycotts were totally coordinated and successfully implemented, their impact on tropical forest depletion would be bound to be small. Only about 6 percent of the total amount of wood cut for all purposes in the tropics enters the international timber trade. Moreover, most of the deforestation taking place results not from forestry activities but from uncontrolled clearing for agriculture, which has social and economic roots.

Even if all international trade in tropical timber were halted, deforestation would continue virtually unabated. Action is needed for the social and economic betterment of the rural people of developing countries, along with the promotion of sound forest management and the infrastructure to support it.

Boycotts or bans could even have a negative effect since, ironically, they are focused upon a commodity that can, in fact, be produced sustainably from tropical forests and that gives the forest a value.[4] If the forest cannot be used for timber exports, the temptation is to clear it and use the land to produce crops, including export crops such as palm oil, sugar, pepper, rice, bananas, coffee or cocoa that are not subject to any threat of boycott.

Moreover, boycotts and bans are directly contrary to international trade agreements such as the General Agreement on Tariffs and Trade (GATT). The Uruguay Round of negotiations on GATT sought to liberalize world trade further in a wide range of goods, including forest products. Furthermore, the forest principles that were developed by consensus at UNCED clearly stated that "unilateral measures, incompatible with international obligations or agreements, to restrict and/or ban international trade in timber or other forest products should be removed or avoided, in order to attain long-term sustainable forest management" (Article 15).

FAO opposes such trade restrictions. FAO believes that these measures are unlikely to make any positive contribution to solving the problems facing tropical forests. Bans on trade in tropical timbers would deprive a country and its rural people of the opportunity to derive economic benefits from an important natural resource. The loss of revenue would only aggravate the very problems that the proponents of bans or boycotts seek to solve. What is needed is material support and encouragement to manage and use forests soundly and sustainably.

[1]In several countries, mainly in Asia, TFAP-type programmes have been launched under the Forestry Sector Master Plan scheme sponsored jointly by the Finnish Department of International Development Cooperation (FINNIDA) and the Asian Development Bank. The principles are the same as for TFAP. Other countries are preparing Environmental Action Plans, often sponsored by the World Bank, within which forest conservation programmes may be included.

[2]"Major coordinating mechanisms established by the UN system following UNCED, include a high-level inter-governmental forum which cuts across all segments and sectors of the United Nations programmes, the Commission on Sustainable Development (CSD). The Secretariat to this 53-member Commission is provided by the newly established Department on Policy Coordination and Sustainable Development. The CSD has a mandate to enhance international cooperation; help rationalize inter-governmental decision-making capacity for the integration of environment and development issues; monitor and review progress related to implementation of Agenda 21 at the international, regional and national levels; review adequacy of financial resources/mechanisms and transfer of technology required for such implementation; interact with other UN bodies dealing with environment and development, and consider information from governments and from international and non-governmental organizations; and provide reports and make recommendations to the General Assembly, through the Economic and Social Council. With regard to representation from non-governmental organizations it has been decided that NGOs with consultative status with UN's Economic and Social Council (ECOSOC) may designate representatives (observers) to the CSD. An Inter-Agency Committee on Sustainable Development (IACSD), chaired by the Secretary-General of the UN and composed of members from nine UN Specialized Agencies, has also been set up to facilitate and gather information on inter-agency cooperation, and to ensure coordinated, system-wide implementation of Agenda 21" (Palmberg-Lerche, 1993).

[3]The following species have been included in the appendices of CITES: Appendix I (trade prohibited): *Abies guatemalensis*, *Araucaria araucana*, *Dalbergia nigra*, *Fitzroya cupressoides*, *Pilgerodendron uviferum*, *Podocarpus parlatorei*; Appendix II (trade controlled through permits): *A. araucana*, *Guaiacum officinale*, *Guaiacum sanctum*, *Oreomunnea pterocarpa*, *Pericopsis elata*, *Platymiscium pleiostachyum*, *Swietenia macrophylla*, *Swietenia mahagoni*, *Swietenia humilis*. One or more geographically separate populations or subspecies of *A. araucana* are excluded from Appendix I or II.

[4]"The Committee, without questioning the intentions of conservation groups advocating a ban on the import of tropical timber by industrialized countries, unanimously and strongly advised against such measures, as they would not contribute towards the objectives of tropical forest conservation. In fact the loss of commercial value of tropical forests in combination with increased unemployment would make forests more exposed to irrational land clearance for subsistence agriculture" (FAO, 1989).

CHAPTER 8

The way ahead

Management of the world's patrimony of forests in a rational
and sustainable manner is one of the critical challenges facing the human race.
It is essential to be realistic in tackling it.

The pressures leading to the loss of forests originate mainly from outside
the forestry sector, and they are certain to continue in the medium-term
future (see e.g. FAO, 1988b). The immediate need is for
constructive interventions and support for actions that will reduce damage
while laying the foundations for the stabilization and ultimate
sustainable management of the remaining forests. Enough of such interventions
have already been identified to provide an agenda
for immediate and urgent action.

The broader development perspective

Forest depletion in the developing world is not fundamentally rooted in logging or even in clearing for agriculture. It is rooted in poverty, underdevelopment and population growth. It is the success in confronting these challenges that will ultimately determine the fate of the greater part of the world's forests.

Condemnation of developing countries for the way in which they are exploiting their forest resources is futile. The processes of encroachment and forest clearing are for all practical purposes unstoppable at the present levels of economic development in the great majority of the tropical countries. The best that can be expected, and it is well worth striving for, is that the processes be managed in a way that makes them less harmful and self-defeating by maintaining a sufficient presence of trees and woodland integrated within farming systems. At the same time, the forests of the tropics must increasingly be brought under effective management to reach the point where they constitute a sound land-use option that provides income for local people and is economically sustainable.

Industrialized countries have no grounds for moral superiority in environmental matters. They remain primarily responsible for ozone depletion, the threat of global warming, most of the use of irreplaceable fossil fuel resources and the depletion of fish stocks and other biological resources. In their climb to their present level of development, their own forest resources have been cleared and exploited as social and economic necessity dictated.

The industrialized countries cannot now expect to pull up the ladder by which they themselves have climbed to prosperity. In urging developing countries to conserve forest resources, they must be sensitive to charges of eco-colonialism or eco-bullying and must accept that they too not only must behave responsibly but also must bear a fair share of the costs of conserving the global environment.

It is also essential to recognize the fundamental differences in socio-economic conditions between the industrialized and developing worlds. The industrialized world has reduced its rate of population growth; it is able to feed its people from a restricted area of sustainably managed land; it is able to provide the basic necessities of clean water, health care and education; it has secure and abundant energy supplies. Many of the industrialized countries have learned how to conserve and utilize their forest resources for sustainable yield of forest products (although there may still be some way to go before fully sustainable forest management is in place). All of these advances, which are taken for granted by the citizens of the industrialized world, are priority targets for the governments of most developing countries. For the majority of people living in the poorest countries, they are still distant dreams.

Clearing the forests will continue as long as people in the developing countries need to find new land to keep themselves from hunger. People and countries that cannot afford the transition to other fuels will continue to burn fuelwood and charcoal. The use of renewable sources of energy as such is to be encouraged; what is harmful is the lack of management of the resource. The same financial and economic pressures that drive the resource exploitation and export industries of the industrialized world will ensure that countries with forest resources will use them for their own internal use or to generate export earnings. The problem is to avoid exploitation of the forest resource and to ensure sound management.

Rising populations contribute to the pressure on the forest lands of the tropics. The world population is now 5 400 million. The forecast is for 6 250 million by the end of this century and 8 500 million by 2025. The major part of this increase will occur in the poorer countries and among the poorest people within them.

While population growth is obviously a cause of much of the pressure on forest lands, it also has to be seen as a symptom of the underlying economic and social position in which people find themselves. When educational and living standards rise the numbers of children decline; as an illustration, women with secondary education in Brazil have an average of 2.5 children while those with little education have an average of 6.5.

Family planning programmes have an obvious role. Many women want to have fewer children but are deprived of the means of doing so; family planning programmes will provide the means. But simply providing the means to limit families will not persuade people to do so unless they are convinced that this makes sense in the economic and social conditions in which they find themselves

In today's increasingly crowded and polluted world, no country can ignore what is happening elsewhere, particularly on or near its borders. The common future of humanity requires that countries act together. This requires understanding, compromise and mutual respect and assistance. Helping the countries of the developing world to make the demographic and economic transitions that will enable them to conserve and manage their forests is in everyone's interests, for the benefit of their own populations as well as the whole of humanity.

An agenda for action

The following actions and policy changes are all within the immediate power of governments and the sphere of the international agencies. If promoted in all zones they should ultimately lead to the sustainable, multipurpose management of natural forests and plantations, the rehabilitation of degraded forests and reforestation. They do not ask for present sacrifices to ensure benefits in the distant future; they make sense in the short term as well as in the long term.

Implementing these actions would do a great deal to defend the forests against the pressures upon them now. It would also set policy in the right direction to lead to stabilization, management and protection of the world's forest heritage.

Forestry in land use
• Harmonize regulations covering all forms of land use (especially forestry and agriculture) and the environment to ensure consistency in the move towards sustainability.

• Ensure that the conversion of forest land to agriculture is done within the framework of a land-use plan and that land converted to farming is suitable for sustainable agriculture.
• Develop integrated and sustainable land-use systems. Train extension service staff to promote the maintenance or establishment of trees and woodlots within farmland as part of sustainable agricultural systems.

Rural development
• Increase agricultural productivity with appropriate technology to raise food production with less dependence on horizontal expansion at the expense of forests.
• Encourage off-farm sources of income and employment in rural areas, giving special attention to the landless and rural poor as beneficiaries of these opportunities.

Institutional action
• Develop forest policies that promote sustainable forest management in the broadest sense.
• Reform forestry legislation and regulations to provide a consistent and comprehensive framework for the long-term sustainability of forests and for the participation of people who depend on them for their management.
• Strengthen forestry services and staff capabilities (by training and

motivation of staff through incentives, reorganization, etc. as appropriate) to provide advice and support for the implementation of sustainable management programmes that are economically feasible, socially acceptable and environmentally sound.

• Promote collaboration and coordination and a multidisciplinary approach among institutions involved in all aspects of land use.

• Provide training to all involved in rural development (including foresters and agricultural staff) in participatory planning and management, with emphasis on the linkages between forest management and sustainable development.

Planning

• Support the drawing up and implementation of forest action plans in all countries, with due emphasis on the sustainable management of natural forests and the systematic integration of actions to maintain biodiversity and to ensure forest protection.

Forest conservation through management

• Encourage the creation and protection of conservation areas in forest ecosystems, especially primary forests.

• Carry out research into technical, social and economic aspects of sustainable forest management. Special attention should be devoted to developing countries and particularly to traditional methods, with emphasis on research into the implementation of participatory approaches, the identification of user needs, methods for the valuation of all forest goods and services and environmentally appropriate techniques for tree planting and harvesting.

• Promote systems of harvesting that ensure the continued supply of forest goods and services and maintain biodiversity.

• Promote long-term management contracts in public forests to enforce environmentally sound forestry practices and to develop partnership arrangements involving local communities, concessionaires, traders in forest products and other interest groups.

• Encourage the broad participation of rural people in drawing up and implementing forest management plans for common property resources, and provide adequate incentives and technical support to encourage collective responsibility and initiatives by local groups.

• Promote the regeneration and management of degraded forest lands in collaboration with local communities to meet their needs and goals, providing assistance and training where required.

All these activities require information, technical skills, political commitment and funding. It is up to the industrialized world to ensure that the burden of providing what is required does not rest entirely on the shoulders of the developing countries. Securing the future of the world's forests is a truly global task in which humanity as a whole must collaborate if progress is to be made and success is eventually to be achieved.

Annex 1

Officially sanctioned and funded debt-for-nature swaps to date
with commercial bank debt (as of December 1992)

Date	Country	Purchaser/fundraiser	Cost (US$)	Face value of debt (US$)	Conservation funds generated (US$)[a]
6/92	Bolivia	TNC/WWF/JP MORGAN	0	11 500 000	2 800 000
6/92	Brazil	TNC	748 000	2 200 000	2 200 000
1/92	Philippines	WWF	5 000 000	10 150 000[b]	9 332 354
10/91	Guatemala/CABEI	TNC	75 000	100 000	90 000
10/91	Jamaica	TNC/USAID/PRCT	300 000	437 000	437 000
7/91	Nigeria	NCF	64 788	149 800	93 446
4/91	Mexico	CI	180 000	250 000	250 000[c]
1/91	Madagascar	CI	59 377	118 754	118 754[c]
1/91	Costa Rica/CABEI	RA/MCL/INC	360 000	600 000	540 000
8/90	Madagascar	WWF	445 891	919 363	919 363
8/90	Philippines	WWF	438 750	900 000	900 000
3/90	Costa Rica	Sweden/WWF/INC	1 953 474	10 753 631	9 602 904
3/90	Dominican Rep.	PRCT/INC	116 400	582 000	582 000
1/90	Poland	WWF	11 500	50 000	50 000
8/89	Zambia	WWF	454 000	2 270 000	2 270 000
7/89	Madagascar	WWF	950 000	2 111 112	2 111 112
4/89	Ecuador	WWF/TNC/MBG	1 068 750	9 000 000	9 000 000
4/89	Costa Rica	Sweden	3 500 000	24 500 000	17 100 000
1/89	Costa Rica	TNC	784 000	5 600 000	1 680 000
1/89	Philippines	WWF	200 000	390 000	390 000
7/88	Costa Rica	Holland	5 000 000	33 000 000	9 900 000
2/88	Costa Rica	NPF	918 000	5 400 000	4 050 000
12/87	Ecuador	WWF	354 000	1 000 000	1 000 000
8/87	Bolivia	CI	100 000	650 000	250 000
Total to date			23 081 930	122 631 660	75 666 933

[a] In the case of bonds this figure does not include interest earned over the life of the bonds.
[b] Included 200 000 debt donation from the Bank of Tokyo.
[c] Debt exchanged for cash, not conservation bonds.
Abbreviations: TNC, The Nature Conservancy; PRCT, Puerto Rico Conservation Trust; NPF, National Park Foundation of Costa Rica; CABEI, Central American Bank for Economic Integration; RA, Rainforest Alliance; WWF, World Wide Fund for Nature; MBG, Missouri Botanical Garden; CI, Conservation International; NCF, Nigerian Conservation Foundation; MCL, Monteverde Conservation League.
Source: The Nature Conservancy, Arlington, VA, USA, personal communication, March 1993.

References

Anderson, A. B. & Posey, D. A.
1989. Management of a tropical scrub savanna by the Korotire Kayapó of Brazil. *Advances in Economic Botany*, 7: 159–173. [Quoted in FAO, 1993. *Sustainable management of forests in the tropics and subtropics for non-wood forest products*. FAO Forestry Paper. Rome. (In preparation).]

Bramble, B. J.
1987. The debt crisis: the opportunities. *Ecologist*, 17(4/5): 192–199.

Brown, L. R. *et al.*
1991. *State of the World 1991*. Washington, DC, USA, Worldwatch Institute.

Cabellero Deloya, M.
1993. Urban forestry in Mexico City. *Unasylva*, 173: 28–32.

Ciesla, W. M.
1993. What is happening to the neem in the Sahel? *Unasylva*, 172: 45–51.

Dembner, S.
1993. Urban forestry in Beijing. *Unasylva*, 173: 13–18.

Dudley, N.
1992. *Forests in trouble: a review of the status of temperate forests worldwide*. 1992. Gland, Switzerland, WWF International.

Economist
1992. *A survey of the global environment.* Insert to Vol. 323, No. 7761 (30 May).

FAO
1988a. *Manual on mapping and inventory of mangroves.* FO:MISC/88/1. Rome.

FAO
1988b. *World agriculture: toward 2000*, ed. N. Alexandratos. 1988. Rome.

FAO
1989. *Report of the Committee on Forest Development in the Tropics, Ninth Session.* Rome.

FAO
1990a. *Climate change and global forests: current knowledge of potential effects, adaptation and mitigation options.* FO:MISC/90/7. Rome.

FAO
1990b. Final report of an FAO/UNEP project in Peru, FAO/UNEP FP 6102-85-01.

FAO
1991a. *Strategies for the establishment of a network of in situ conservation areas.* Secretariat Note CPGR/91/6. Fourth Session of the FAO Commission on Plant Genetic Resources, Rome, 15–19 April

1991. [Published in FAO, 1992. *FAO activities on* in situ *conservation of plant genetic resources.* Forest Genetic Resources Information No. 19. Rome.]

FAO
1991b. *Report of the Committee on Forest Development in the Tropics, Tenth Session.* FO:FDT/91/REP. Rome.

FAO
1991c. *Sustainable management of tropical forests.* Secretariat Note. FAO Committee on Forest Development in the Tropics, Tenth Session. FO:FDT/91/5. Rome.

FAO
1992a. *FAO yearbook of forest products 1990.* FAO Forestry Series No. 25/FAO Statistics Series No. 103. Rome.

FAO
1992b. *Report on forest resources assessment 1990.* Paper to the FAO Committee on Forestry, Eleventh Session, November 1992. COFO-93/2

FAO
1993a. *Cosecha de hongos en la VII región de Chile.* Estudio Monográfico de Explotación Forestal, 2. Rome.

FAO
1993b. *Conservation of genetic resources in tropical forest management: principles and concepts.* FAO Forestry Paper 107. Rome.

FAO
1993c. *Forest resources assessment 1990 – tropical countries.* FAO Rome.

FAO
1993d. *Sustainable management of tropical moist forest for wood.* [To be published in *The Challenge of Sustainable Forest Management: Technical Papers.* FAO Rome. (In preparation)]

FAO
1993e. *Policy, legal and institutional aspects of sustainable forest management.* [To be published in *The Challenge of Sustainable Forest Management: Technical Papers.* FAO Rome. (In preparation)]

FAO
1993f. *Forest decline: a global perspective.* FAO Forestry Paper. Rome. (In preparation).

FAO
1993g. *Ensuring sustainable management of wildlife resources: the case of Africa.* [To be published in *The Challenge of Sustainable Forest Management: Technical Papers.* FAO Rome. (In preparation)]

FAO
1993h. *Climate change and sustainable forest management.* [To be published in *The Challenge of Sustainable Forest Management: Technical Papers.* FAO Rome. (In preparation)]

FAO
1993i. *Mangrove forest management guidelines.* FAO Forestry Paper. Rome. (In preparation).

FAO
1993j. *Sustainable management of plantation forest in the tropics and subtropics.* [To be published in *The Challenge of Sustainable Forest Management: Technical Papers.* FAO Rome. (In preparation)]

Gauthier, J. J.
1991. *Plantation wood in international trade.* Paper presented to the seminar Issues Dialogue on Tree Plantations – Benefits and Drawbacks, Centre for Applied Studies in International Negotiations (CASIN), Geneva, April 1991.

Global Environmental Change Report
1991. Policy trends: New York State passes tropical timber ban. 3(16): 3.

Hoffman, M. S., ed.
1992. *World almanac and book of facts.* New York, Pharos Books.

Intergovernmental Panel on Climate Change (IPCC)
1990. *Policymakers' summary of the formulation of response strategies.* Report prepared for IPCC by Working Group III, June 1990.

Koch, P.
1991. *Wood vs non-wood materials in US residential construction: some energy-related international implications.* Working Paper 36. Seattle, WA, USA, Center for International Trade in Forest Products, University of Washington. [Quoted in H. Salwasser, D. W. MacCleary & T. A. Snellgrove. 1992. New perspectives for managing the US national forest system. *North American Forestry Commission, Sixteenth Session, Mexico, February 1992.* Rome, FAO.]

Kuchelmeister, G. & Braatz, S.
1993. *Urban forestry revisited.* Unasylva, 173: 3–12.

Lanly, J. P.
1992. Forestry issues at the United Nations Conference on Environment and Development. *Unasylva,* 171: 61–67.

Lanly, J. P. & Allan, T.
1991. Overview of status and trends of
world's forests. In *Proceedings of the Technical
Workshop to Explore Options for Global
Forestry Management*, Bangkok, 1991.
Bangkok, Office of the National
Environment Board/UK, International
Institute for Environment and
Development/Japan, ITTO.

Ledig, F. T.
1986. Conservation strategies for forest
gene resources. *Forest Ecology and
Management*, 14: 77–90.

Noordwijk Ministerial Conference
1989. *The Noordwijk Declaration on Climate
Change*. Ministerial Conference on
Atmospheric Pollution and Climate
Change, Noordwijk, The Netherlands,
November 1989. The Netherlands,
Government of the Netherlands.

Nowak, D.J. & McPherson, E.G.
1993. Quantifying the impact of trees: the
Chicago Urban Forest Climate Project.
Unasylva, 173: 39-44.

Palmberg, C.
1987. Conservación de recursos genéticos
de especies leñosas. *Actas, Simposio sobre
Silvicultura y Mejoramiento Genético de Especies
Forestales*, 11: 58–80. Buenos Aires, Centro
de Investigaciones y Experiencias Forestales.

Palmberg-Lerche, C.
1993. *International programmes for the
conservation of forest genetic resources*. Invited
paper for the International Symposium on
Genetic Conservation and Production of
Tropical Forest Seed, Chiang Mai, Thailand,
June 1993. Jakarta, ASEAN/Canada Forest
Tree Seed Centre, in collaboration with the
Government of Thailand, FAO, the Forestry
and Fuelwood Research and Development
Project and the International Development
Research Centre.

Pandey, D.
1992. *Assessment of tropical forest plantation
resources*. Institutionen för Skogtaxering,
Swedish University of Agricultural Sciences.

Reid, W. V. *et al.*
1993. *Biodiversity prospecting: strategies for
sharing benefits*. Paper to the International
Conference on the Convention on
Biological Diversity: National Interests and
Global Imperatives, UNEP, Nairobi,
January 1993.

Schroeder, P.
1991. *Carbon storage potential of short rotation
tropical tree plantations*. Corvallis, OR, USA,
United States Environmental Protection
Agency.

Sedjo, R. & Solomon, A.
1989. *Climate and forests.* Paper prepared for workshop on controlling and adapting to greenhouse forcing, 14–15 July 1988. Washington, DC, USA, United States Environmental Protection Agency, National Academy of Sciences.

Swedish Forest Service
1978. *Domän Posten*, Vol. 182 (September).

Thirgood, J. V. 1981
Man and the Mediterranean forest: a history of resource depletion. London, Academic Press.

Thirgood, J. V. 1989
Man's impact on the forests of Europe. *Journal of World Forest Resource Management*, 4(2): 127–167.

United Nations Conference on Environment and Development (UNCED)
1992. Agenda 21, Chapter 11: Combating deforestation.

Whitmore, T.C.
1990. *Tropical rain forests.* Oxford, UK, Clarendon Press.

Wilcox, B.A.
1984. Concepts in conservation biology: applications to the management of biological diversity. In J. C. Cooley & J. H. Cooley, eds. *Natural diversity in forest ecosystems: proceedings of the workshop*, p. 155–172. Athens, GA, USA, University of Georgia.

World Bank.
1991. The forest sector. *A World Bank policy paper.* Washington, DC.

World Commission on Environment and Development
1987. *Our common future.* New York, Oxford University Press.

World Resources Institute
1993. *Biodiversity prospecting: using genetic resources for sustainable development.* Baltimore, MD, USA, World Resources Institute.